IN THE SHADOW OF PROGRESS

Eric Cohen

IN THE SHADOW OF PROGRESS

Being Human in the Age of Technology

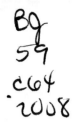

New Atlantis Books

ENCOUNTER BOOKS · NEW YORK · LONDON

First edition published in 2008 by Encounter Books,
an activity of Encounter for Culture and Education, Inc.,
a nonprofit, tax exempt corporation.
Encounter Books website address: www.encounterbooks.com

Manufactured in the United States and printed on acid-free paper.
The paper used in this publication meets the minimum requirements of
ANSI/NISO Z39.48-1992 (R 1997) (Permanence of Paper).

FIRST EDITION

LIBRARY OF CONGRESS CATALOGING-IN-PUBLICATION DATA

Cohen, Eric, 1977–
In the shadow of progress : being human in the age of technology / Eric Cohen. —
1st hardcover ed.
    p.   cm. — (New Atlantis books)
Includes bibliographical references and index.
ISBN-13: 978-1-59403-208-0 (hardcover : alk. paper)
ISBN-10: 1-59403-208-4 (hardcover : alk. paper)
1. Technology—Moral and ethical aspects. I. Title.
BJ59.C595   2008
174′.96—dc22
2008001919

10   9   8   7   6   5   4   3   2   1

CONTENTS

TO MY PARENTS,

JEFFREY AND MAUREEN COHEN

The Question of Progress

WHEN MY SON was born, he had trouble breathing—visible to his nervous father in the bluish tint of his skin, diagnosed by the attending physician from the low percentage of oxygen in his blood. So off he went, within minutes, to the neonatal intensive care unit (NICU). He was, truth to tell, the healthiest child in a row of babies struggling to survive. He was a giant among "preemies"—there for observation, not because only modern machines and master physicians could keep him alive. After a few days, he was released and reunited with his waiting mother. We rolled him from the third floor to the sixth, from NICU to normal, from anxiety to relief. Blessed by nature, blessed by science, we left the hospital "on schedule." We also knew that many children did not, and that some children never left at all. Nature is not always so generous; science is not always so powerful. But for us, in those moments, we marveled at both.

A few days later, my son was circumcised in the traditional Jewish way—in our home, by a mohel, with grandparents present. The mohel's instruments were archaic compared to the machinery of the NICU; the modern beeping of the incubators gave way to ancient blessings and rituals that have persisted from generation to generation for

1

millennia. If the NICU exemplifies the most elevated ambitions of progress—to heal the broken body, to sustain life in the face of death, to correct nature's mistakes—the bris is a reminder of the permanent horizons of human life—the mystery of human origins, the cycle of the generations, the eternal drama of rearing those who will one day stand in one's place, and, in the normal course of things, at one's grave-side. The bris embodies the eternal hope that lies beyond progress; what it aims to sanctify—the welcoming of a new child—is not a novel artifact of the current age but a primordial experience of being human.

Yet, remarkably, one of the defining features of the modern era is that the most modern individuals are not having enough children to sustain their societies from one generation to the next. Communities defined by their ancient faith continue to have children in high num-bers, believing they have something sacred to sustain in the flesh and rearing of their young. But those most immersed in the pleasures and possibilities of modern life seem least driven to raise up a generation to follow in their footsteps. Societies defined by the forward march of progress are failing to bring life forward in the most fundamental sense. What faith, we are left to wonder, does modern man have in the cosmic significance of modern, individualist, technological life? If procreation is the deepest form of fidelity to one's civilization, then what does modern man's infidelity say about the relative greatness (or goodness) of the modern age? Is progress really progress?

One might ask this question—is progress really progress?—about many domains of modern life: Is it progress to ameliorate human sad-ness by altering the chemical balance of the brain? Is it progress to seek cures for the sick by destroying human embryos? Is it progress to screen and select the genetic make-up of one's offspring? Is it progress to wield the force of the split atom, or to transform the habitats of nature into oil fields?

To question progress is a perilous enterprise, since those of us who are in a position to question it are typically its most fortunate beneficiaries.

Blessed as we are to live in the age of modern science, gratitude is the only authentic response—the gratitude of the parent whose child is cured of cancer, the gratitude of the wife who can hear her husband's voice from distant battlefields, the gratitude of the son who flies home in time to say one final goodbye to his father. Ancient accounts of human experience—David's longing for salvation, Homer's tales of honor, Shakespeare's portraits of love—still speak to us. But living, as we do, in the age of vaccines and painkillers and running water, we can never fully fathom the material conditions of life that confronted the men, women, and children of the past, whose lives were often nasty, brutish, and short.

Yet the current age, for all its great blessings, leaves much room for disquiet and many human longings unsatisfied. The most obvious source of our disquiet is the fear of technology's power to wreak havoc, and our inability to control the dark uses of our own inventions. If endless improvements in modern medicine are highly likely, so is the proliferation of modern weapons. If modern life, in general, affirms our faith in progress, the political realities of the age point toward the possibility of killing on a mass scale, with novel instruments of destruction that we can barely now imagine. Gratitude and terror, endless improvements and looming disaster, the twin expectations of better health via stem cells and gruesome death via biological weapons—this is our technological condition, seen in its starkest form.

For most people, however, everyday life is not dominated by such utopian dreams or apocalyptic nightmares. The very comfort made possible by the arts of progress puts greater distance between us and these extremes, and it is precisely because we have achieved so much in softening the harsh edges of the human condition that we struggle to live well with the limits of progress. The cardiologist can often heal the sickly heart but never the broken heart. And the neonatologist, for all his successes, is also haunted by the tiny bodies who breathe their last breath on his watch. For the children of the NICU that never go home, the observant Jew mourns in the traditional way, by saying Kaddish. The concrete, inexplicable tragedy of this dead child is assimilated into the larger mystery of mortality and the permanent hope of redemption. But from the perspective of science, this death is not a mystery, or even a tragedy, but a failure. To believe only in progress is to see this failure—the victory of death against man's efforts to oppose it—as our only destiny. To say Kaddish, and to circumcise the young, is to believe otherwise.

This book aims, with some trepidation, to lean against America's faith in progress—not out of ingratitude for the tremendous blessings of the modern age, and not solely out of fear that modern technology may be used in the service of mass death. Instead, the book questions progress by exploring its incompleteness as an answer to man's deepest longings—the longing to be loved, the longing to be virtuous, the longing to be redeemed. It also explores the dangers of becoming corrupt, or flat-souled, in the unbridled pursuit of a life without sickness, hardship, shame, or death.

For when it comes to modern technology, the sources of our greatest pride—especially our growing power to cure disease and extend life—clearly bring novel perils. The search for solutions to man's problems—especially the problems of the body—tempts us to transgress important ethical boundaries in the quest to make life better. The history of science is filled with many awful experiments that were purportedly justified by their good intentions; and the boldest practitioners of modern biological science, while decrying the terrible experiments of the past, seem all too willing to embark on questionable experiments of their own.

In recent years, this moral drama has played out most passionately in the battles over embryo research—the embryos we would destroy in the name of saving the loved ones we cannot bear to lose. We would not love if we did not hunger for such cures; yet in the name of love, science blinds us to our most basic moral obligations: never to use some lives as tools to help others, never to dehumanize the weak in the name of helping the strong. The disaggregated embryo is the hidden face of progress we choose not to see; its severed cells embody the peril that comes with seeking too desperately to remedy the ills of mortal life by human artfulness alone.

At the same time, technology affords us numerous avenues for removing ourselves from the complicated immediacy of life—with its ups and downs, joys and griefs, burdens and responsibilities. Growing numbers of people are living imaginary lives in the virtual world, creating bodiless identities where they can live out their private fantasies, large and small. Meanwhile, ever more people are turning to mood-altering drugs, hoping to numb themselves, at least partially, from the inevitable discontents of being human. From the stem cell laboratory to our primitive versions of Huxley's "feelies" and "soma," biotechnology invites us to believe that we can conquer mortal life while also affording us novel means to retreat from mortal life. It both inflates and deflates our human expectations simultaneously.

Of course, facing up to the moral challenges of progress should never obscure either its blessings or its necessity. Science and technology are necessary to defend ourselves against the dark plans of civilization's most determined enemies. In this sense, whether we like it or not, we are bound to the technology project. And the redeeming goods of life —having children, beholding nature's beauty, passing down a religious tradition—are often only possible because of our technological interventions against the blind entropy of physical nature. The achievements of science are among the noblest testimonials to the special dignity of man, the creature with wonder, the creature who refuses to lie impotent in the face of blind nature, the creature who seeks to understand the hidden workings of the world. If one could choose when to be born, the present age surely makes a compelling case, arguably the most compelling case. Modern terrors—like nuclear war—should not blind us to the terrors of the past—like the black plague—that left men naked unto death.

Yet it is also misguided to believe that the sages of the past have nothing to teach us, or that our world is necessarily the summit of human excellence, or that the present age is without blemishes. To believe that progress is real, as I surely do, does not mean accepting that every aspect of human life is improving. At the heights, we can wonder whether we will ever produce music as sublime as Mozart's, or art as marvelous as Michelangelo's, or sacred texts as elevating as the psalms, or statesmen as great as Churchill. In myriad realms where man displays who he is at his best—art, virtue, wisdom—the towering achievements of the past still command our awestruck admiration. At the level of everyday life, we can wonder whether the civilization that chooses "the Pill" over the bris really believes in the future, or whether the culture that turns increasingly to anti-depressants is truly happy.

Taken together, the aim of this book is to think soberly about progress —by exploring when to spread its fruits and when to limit its excesses, by grappling with both its superiority and its deficiencies, by romanticizing neither the lost worlds of the past nor the imagined worlds of the future. Our challenge is to live simultaneously with gratitude and fear, pride and shame, sobriety and hope, in this new age of technology. This book aims to articulate that challenge, and to gesture toward some ways of dealing with it.

The first section explores the aspirations and limits of the modern scientific project. It aims to see various contemporary developments—from stem cells, to evolutionary theory, to space travel, to man-animal experiments—in light of the original aims and philosophical foundations of modern science. And it dissects the scientific image of man—seen as both the godlike ruler of nature and the beastlike product of nature—in an effort to recover a richer understanding of human life and human longing.

The second section explores, in a more focused way, the ethics of progress—and especially the moral dilemmas set before us by our growing powers over birth and death, body and psyche, mood and memory. It explores how rapid advances in biotechnology will put the American character on trial, calling into question the meaning of America's foundational commitment to human equality.

The final section aims to situate the dilemmas of progress within the eternal drama of birth and death, permanence and change, finitude and eternity. It invites us to ask: Why have children? What does it mean to age? In whose image shall we die? A civilization with poor answers to these questions, or poor ways of dealing with them, can take only limited pride in its unlimited improvements, especially if it forgoes the true hope of every generation, which is the continuity of life in the flesh of the young.

If this were the seventeenth century, I hope I would have been warily on the side of Bacon and Descartes, moved by human charity to be a partisan of science in the relief of man's estate. Back then, they were the contrarians. Today, however, being a contrarian means being willing to ask what we have lost in an age that takes the virtues of progress for granted. In doing so, we are in good American company. Nathaniel Hawthorne, arguably America's greatest literary figure, wrote tales that mocked man's illusions of godlike power over nature and human nature. Original sin, in Hawthorne's vision, always found its revenge; our boldest attempts to eradicate the marks of our birth and the perversions of our soul end in disaster or, at best, prove mixed blessings. We should heed Hawthorne's warning even as we celebrate our achievements. For the trials of progress are great, and it is hard to believe, in the end, that modernity will prove its perfect innocence in the eyes of eternity.

PART ONE

Science and the Human Prospect

CHAPTER ONE

The Spirit of Modern Science

WHENEVER I SPEND time with scientists, I'm always struck by their optimism—and their discontent.

Mostly they are optimists, excited by the latest findings: the newly isolated gene variant that may help explain schizophrenia, the new telescopic images that reveal the violent births of distant galaxies, the geochemical discoveries that may change our understanding of Earth's formation. Armed with an endless array of PowerPoint slides, the optimists believe they are uncovering life's secrets slice by slice, defining humanity's place in the universe, making life better through their mastery of nature's mechanisms. Knowledge through experiment, progress through reason: They have no doubt they are on the right side of history.

And yet, at the same time, many of these scientists seem frustrated and unhappy. Some are furious because policymakers are ignoring their advice and trying to police their laboratories. Some believe that religious fundamentalists are on the march, replacing the study of evolution with the pseudoscience of intelligent design. Some fear that human beings are poised to wreck the natural world by polluting the atmosphere or poisoning the soil. Others feel defeated by nature's relentless brutality, by the tsunamis and earthquakes and childhood cancers that so regularly mock man's illusions of control. "Don't blame God," a caption in *Science* magazine exhorts us, alongside a snapshot of the devastation from the 2004 tsunami in Southeast Asia. "Better planning could make natural disasters much less disastrous."[1] But that hardly seems to mitigate nature's relentless indifference to young and old alike, or ameliorate the misery of the helpless mothers who were left to mourn the dead infants of Java Island.

Perhaps one reason the debate about embryonic stem cells has become so prominent in recent years is that it combines scientific optimism and scientific despair so completely: the optimistic search for cures, the discontent that nature yields remedies for her afflictions so slowly, the frustration with "anti-science" moralists who seek to stand in the way of scientific progress for nonscientific reasons. The deans of major research centers feel like persecuted Galileos, even as they defend their turf in the most unscientific ways: promising cures based only on preliminary experiments, caricaturing their opponents as irrational fanatics, making emotional appeals designed to obscure what embryos truly are and what obligations we have toward them.

The embryo debate is suggestive of a larger problem: the methods of science cannot vindicate the ends of science, and the knowledge acquired by scientific methods cannot always justify the particular experiments used to acquire it. Yet scientists desperately want such vindication in the eyes of their fellow citizens: Good science (meaning *interesting, promising, exciting*) needs to be seen as good (meaning *virtuous, praiseworthy, compassionate*) by everyone. And so scientists have invented a new method to defend the unfettered freedom of the old one: They claim the mantle of science while making ethical claims ("embryo research is morally good") that rest on no special scientific basis at all, and they portray their opponents as anti-science for raising ethical questions that are entirely consistent with the scientific facts ("embryological development begins at conception").

Our current controversies about embryo research or human cloning or evolution are not simple Manichean struggles between science and religion, contrary to the pop-culture perception. In these battles, the relationship between science and religion is far more complex, more ambiguous, more puzzling. Many scientists surely see all religious faith as an illusion. They dismiss all limits on science as based in faith alone and therefore without rational authority. Yet the scientists themselves are often moved by a quasi-religious purpose—the desire to be redeemers of the body—that has no purely scientific ground whatsoever. And while religious believers may, at times, overstate (or even distort) the scientific case for their particular views about the creation and destiny of man, they also make many moral arguments that rest on sound scientific premises—far more sound, in fact, than the fuzzy falsehoods that many scientists sometimes use to advance the cause of science, like declaring that embryos are just a "ball of cells," no different than a bundle of materials at Home Depot.[2] One does not need to believe in God or appeal to Scripture to see embryo research as an affront to human equality—an ideal that most scientists embrace, at least in theory, even without adequate scientific proof. And one does not need to believe in the Genesis account of human origins to believe that neo-Darwinian theory fails to address the most interesting questions about human nature and human purpose.

At their best, the devotees of science and of religion both seek the truth about man and nature; both see the human drama as the quest for certainty in the face of mystery; and both long for a future that is better than the present. Where they differ, when they differ, is in the source of man's greatest hope: Is it the possibility of a loving God who remembers His fallen creations, or the possibility of a new science that remakes man the way a loving God should have made him the first place? "The two world-views—science-based explanations and faith-based religions—cannot be reconciled," the esteemed biologist Edward O. Wilson recently wrote. "What then are we to do? Put the differences aside, I say."[3] But such benign yet separate coexistence hardly seems like a viable option. The two realms mix whether we like it or not, and to understand the meaning and limits of modern science, it is helpful to understand, more deeply, the philosophical roots of scientific hostility to religion—a task we can hardly leave to the scientists themselves.

THE NEW SALVATION

Perhaps no idea offends the modern scientific mind more deeply than divine salvation. How weak we must be if we need a God to rescue us from the burdens of living in this world; how foolish we must be to let the fear of offending our imaginary savior get in the way of genuine progress. In *A Devil's Chaplain*, the biologist Richard Dawkins defines religious faith as a "virus of the mind"—or that childlike need to "suck at the pacifier of faith in immortality."[4]

Otherworldly salvation comes in many different guises—salvation for those who believe in the one true God, salvation for those who lead good lives, salvation for those whom God chooses for His own inscrutable reasons. But the basic idea—"The Lord is my light and my salvation; whom shall I fear?"[5]—is the common heritage of Jews and Christians, who yearn for redemption in a world whose many blessings always exist under a cloud of misery. And yet the scientists, empirical to the core, know that faith is a fraud, a delusion, a prison cell for small minds. And it bothers them to the depths of their rational souls—although, curiously, for a series of contradictory reasons.

These contradictions go back to the beginnings of modern science. In *The Great Instauration*, one of the founding texts of the modern age, Francis Bacon described what he believed to be the aim and meaning of human knowledge: "I would address one general admonition to all: that they consider what are the true ends of knowledge, and that they seek it not either for pleasure of mind, or for contention, or for superiority to others ... but for the benefit and use of life, and that they perfect and govern it in charity.... [From the marriage of the Mind and the Universe] there may spring helps to man, and a line and race of inventions that may in some degree subdue and overcome the necessities and miseries of humanity."[6]

The intelligent, in other words, have a duty to their fellow men: to seek knowledge in a way that ameliorates human misery, to seek power in the name of human charity. The trouble, as Bacon knew, is that the beneficiaries of his charity might not always be so amenable to his methods—methods that require violating not only the natural boundaries that exist between the species but also the divine boundaries that long divided the sacred from the profane. Where Leviticus ritually separates pure from impure with an eye to what is divine in man, Bacon's "New Atlantis" vivisects and recombines everything for the sake of healing man's animal body. "We have also parks and enclosures of all sorts of beasts and birds which we use not only for view or rareness, but likewise for dissections and trials; that thereby we may take light what may be wrought upon the body of man," the father of Salomon's House tells his visitors in Bacon's fable.[7] On the isle of progress, the priest is replaced by the scientist, who conducts secret experiments to help his fellow citizens. This is the new charity.

Yet charity alone is hardly the only scientific motivation. For are we really to believe that the scientist's own pleasure is so unimportant to Bacon's project—the pleasure of knowing the world as it really is, taking it apart and putting it back together, coercing nature to reveal her true secrets? Are we really to believe that Descartes, who gave these fabulist visions a mathematical method, was not driven also (or primarily) by new aristocratic pleasures of the mind, pleasures that required the unfettered freedom to experiment?

From the beginning, science was driven by both democratic pity and aristocratic guile, by the promise to help humanity and the desire to be free from the constraints of the common man, with his many myths and superstitions and taboos. The modern scientist comes to heal the wretched bodies of those whose meager minds are always a threat to experimental knowledge. Salomon's House, where the elite of Bacon's scientific utopia would decide which inventions to publish and which to hide, existed both to protect men from science and science from men. It offers a new salvation and seeks to elude the oppressive trappings of the old one. It brings a new compassion and a new contempt. This was true in the beginning, and it is true today.

This double origin of modern science takes another form as well: seeing the unnecessary misery of false hope in a loving God and the untold benefits of tragic realism about blind nature. The spiritual confidence of the believer offends those who know the twisted indifference of nature and know that nature is everything. To the scientist, the believer is filled with more hope than he deserves to be, surprised by a joy that is not real, dreaming of a happy ending that will never come. Callous fate, not divine salvation, is the scientific news—and it is hardly good. As Alfred North Whitehead argued in his 1925 lectures on "Science and the Modern World":

> The pilgrim fathers of the scientific imagination as it exists today are the great tragedians of ancient Athens, Aeschylus, Sophocles, Euripides. Their vision of fate, remorseless and indifferent, urging a tragic incident to its inevitable issue, is the vision possessed by science. Fate in Greek tragedy becomes the order of nature in modern thought. . . . This inevitableness of destiny can only be illustrated in terms of human life by incidents which in fact involve unhappiness. For it is only by them that the futility of escape can be made evident in the drama. This remorseless inevitableness is what pervades scientific thought. The laws of physics are the decrees of fate.[8]

Darwin put this truth in a new, biological light: Man emerged from this "remorseless and indifferent" nature, leaving us with a tragic mismatch between our spiritual longings and our mortal condition.[9] In the Darwinian view, our origins and our destiny are little comfort to those seeking meaning beyond the imminent or seeking redemption from the wretched errors of nature that leave babies to suffer and villains to prosper. Yes, man can take a certain satisfaction and experience a certain "grandeur" in his own natural ascent, as Darwin writes at the end of his masterpiece.[10] But faith that the human story will have a truly comic ending, that it will end in a way that satisfies man's redemptive longings, that Providence is still at work, is weakened if not shattered.

But, of course, this tragic vision of life was not the whole truth about the birth of modern science, with its eyes set from the beginning on lifesaving "invention." "By far the greatest obstacle to the progress of science," wrote Bacon, "is found in this—that men despair and think things impossible."[11] If Christian hope offends the scientist, so does the believer's passive acceptance of misery and active obsession with sin. To live in a godless world means that we are just as innocent as nature is blind—free not to suffer, free to alter nature's workings for our own purposes, free to challenge the cold decree of fate as best we can muster.

Thus Condorcet, the French prophet of man's self-improvement, believed he was living in the "ninth stage" of mankind's progress, when reason will "lift her chains, shake herself free from some of them, and, all the time regaining strength" from the effects of the Christian Dark Ages, to "prepare for and advance the moment of her liberation."[12] As he proclaimed in his *Sketch for a Historical Picture of the Progress of the Human Mind* (1795), "Nature has set no term to the perfection of human faculties; that the perfectibility of man is truly indefinite; and that the progress of this perfectibility, from now onwards independent of any power that might wish to halt it, has no other limit than the duration of the globe upon which nature has cast us."[13]

Not all contemporary scientists are quite as bullish about progress as Condorcet, with his limitless optimism about human destiny within the world rather than beyond it. Some even fear that the "duration of the globe" is quickly coming to an end through man's ecological abuse. But Condorcet's spirit still pervades the modern laboratory, especially the biological laboratory, which is now the high kingdom of empirical science. Once we see (with Darwin) that men are beasts, ascendant in nature but not created in the image of God, we are free to re-create nature as if we were gods. We are liberated to perfect the body and enhance the mind in the scientific image.

But Condorcet's original error—call it the original sin of the scientific Enlightenment—still haunts modern science: Perpetual progress is not the same thing as perfection. Infinite progress also means infinite discontent, as man is left in a state of eternal becoming with no end. "Indefinite perfectibility," Condorcet's dream, is an irreconcilable contradiction.

Perfection, after all, is an end, an ideal, something definite. For Jews, God is the perfect arbiter of justice and mercy. For Christians, Christ embodies the perfection of love. For philosophers, wisdom is the perfect grasp of being in its fullness. Yet the scientist destroys the possibility of perfection by seeing a world in permanent flux. Perhaps the only perfection available to the modern scientist is stoic acceptance of nature's contingency on the way to natural oblivion—and indeed, there is no necessary contradiction between stoic philosophy and modern natural science. Yet stoic acceptance of nature is precisely what modern science, technological from the beginning, is incapable of embracing in spirit. Modern science portrays a world where acceptance of our fate within nature is all we can do, and yet it remakes knowledge in such a way that technological striving against nature is seen as the only thing worth doing. Modern biology, like Sisyphus, is haunted by temporary successes and ultimate failure. It fends off death but cannot eradicate it; it explains death's role in natural selection but not the death of individual men still thirsty for salvation.

Writing just after World War I, when the slaughtered troops and brutal reality of technological war had shattered some of the Enlightenment's optimism, Max Weber describes this tragic aspect of modern science with great pathos:

> For civilized man death has no meaning. It has none because the individual life of civilized man, placed into an infinite "progress," according to its own imminent meaning should never come to an end; for there is always a further step ahead of one who stands in the march of progress. And no man who comes to die stands upon the peak which lies in infinity. . . . He catches only the most minute part of what the life of the spirit

brings forth ever anew, and what he seizes is always something provisional and not definitive, and therefore death for him is a meaningless occurrence. And because death is meaningless, civilized life as such is meaningless; by its very "progressiveness" it gives death the imprint of meaninglessness.[14]

Weber's essay on "science as a vocation" is perhaps the best starting point for understanding the limits of scientific aspiration in our time, and for exploring what it means to live in an age when the impossibility of perfection is a first principle of being. Weber praised scientists for living in the world of facts and criticized those who sought salvation by pretending that the old gods still exist. But he also reminded scientists that they have nothing privileged to say about the realm of value, the realm that matters most to human beings seeking knowledge of how to live. Like everyone else, the scientist must decide which ends to pursue, which gods to serve, which demon will hold "the fibers of his very life."[15] And these are exactly the questions that the scientific method cannot answer. Divine salvation may be an illusion but so is believing that science can tell us how to live in the world it dissects and describes, and how to live well in a world where scientific power is so readily, so seductively, so dangerously at our disposal.

POWER WITHOUT WISDOM

The impotence of science is perhaps most readily apparent in that realm where science is most powerful: nuclear weapons. Consider, for example, the tension over Iran's nuclear program. Only the scientifically trained can provide accurate estimates about the state of Iran's weapons development, about the state of our own offensive and defensive military options, about the likely effects of a nuclear attack—the number of casualties, the hazards to the environment, the technical challenges of rebuilding. But when it comes to making practical decisions or evaluating the meaning of our nuclear predicament, the methods of science lead us into the nonscientific realm of interpretation.

The sociobiologist, who sees man in light of his long animal history and sees nature as a vast impersonal process, might say that the confrontation with Iran hardly matters in the cosmic scheme of things. Man killing man, culture clashing with culture, is inherent to nature's law of survival. Superior cultures, like superior individuals, triumph over inferior ones, at least in the long run. Life improves through death in a blind drama that continues unabated. The neuroscientist, who studies the brain to understand its mechanisms and improve its workings, might look instead to man himself as the agent for changing human nature. Perhaps our advancing understanding of neurobiology will eventually make tyranny a thing of the past. Perhaps our emerging science of the mind will bring a new age of man-made peace and stability, a "psychocivilized society."[16] Perhaps Darwin's greatest species will triumph over Darwin's brutal laws.

Yet such interpretations are morally and strategically unsatisfying, to say the least. Human beings may be destined always to kill one another, but we leave ourselves morally impotent if we see this dark fact about the human condition as our only guide to moral action. Mankind may be destined to become something better, but discerning the difference between self-improvement and self-degradation requires some standard beyond the imminent processes of nature, lest we make ourselves into the subhuman denizens of Huxley's *Brave New World*, living peaceful lives without purpose. And while the unrestrained pursuit of knowledge is perhaps the core dogma of science, one imagines that any scientist with a conscience would reject the shelter of scientific freedom that even a despotic regime like Iran might offer, a regime that might see the uninhibited investigation of the physical world as useful to its own perverse ends. While the *moral* obligation should be obvious, there is no *scientific* reason not to become an Iranian scientist.

In every area of public life where science and morality intersect, there are questions about the use of science that science itself can never answer. On stem cells, scientists can tell us the potential benefits of destroying human embryos but not whether the progress of medicine justifies the willful destruction of nascent human life. On drilling in Alaska, scientists can estimate the potential oil reserves and the potential harm to the ecosystem but not whether our responsibility to expand the domestic oil supply outweighs our responsibility to preserve the unsullied wonders of nature. On the human exploration of space, scientists can estimate the economic and human costs of putting a man on Mars and the potential benefits of such a mission to the advance of human knowledge, but they cannot say whether human greatness in space is more worthy of public funds than ongoing research into curing AIDS. Science is power without wisdom about the uses of power. Science is a means to many ends without wisdom about which ends are most worthy. As the philosopher Hans Jonas put it: "The scientist himself is by his science no more qualified than others to discern, nor is he more disposed to care for, the good of mankind. Benevolence must be called in from the outside to supplement the knowledge acquired through theory: it does not flow from theory itself."[17]

Yet the scientists still often want to tell us how to live, and they often claim the authority of science for their moral exhortations. Richard Dawkins, for example, ends *A Devil's Chaplain* with a letter of advice to his ten-year-old daughter on "good and bad reasons for believing." "Sometimes people have a strong inside feeling that somebody loves them when it is not based upon any evidence, and then they are likely to be completely wrong," he writes. These false feelings pass from one generation to another, from gullible parents to gullible children. "Could this be what has happened with religions"—this perpetuation of illusion? "Belief that there is a god or gods, belief in Heaven, belief that Mary never died, belief that Jesus never had a human father, belief that prayers are answered, belief that wine turns into blood—not one of these beliefs is backed up by any good evidence. Yet millions of people believe them. Perhaps this is because they were told to believe them when they were young enough to believe anything."[18]

One can surely respect the integrity of the rationalist who doubts the existence of a Redeemer he cannot see and who is skeptical about theological claims that rest on ancient texts and dueling authorities rather than empirical evidence. But now imagine, say, a stem-cell biologist writing a similar letter to a ten-year-old girl in the cancer ward— a girl dying of the very disease the biologist cannot yet cure. The girl faces her demise with courage; she knows that God loves her, that the death of her body is not the end of her being. She prays every night, "Even though I walk through the valley of the shadow of death, I will fear no evil, for you are with me," and she still manages to smile every morning.[19] What would the stem-cell biologist say to the girl he wants so desperately to rescue from the ravages of nature? Would he describe the miracle cures that will not come in time? Would he tell her that God's love is an illusion, that her prayers evaporate unheard and unanswered into the ether, that her brief transitory existence is all there is, that she is "sucking the pacifier of faith in immortality"?[20]

Perhaps the rationalist can stomach a little bit of comforting illusion for a dying girl he cannot help. Or perhaps he believes her piety must be shattered for the greater good, since the lives of future children depend on destroying that fundamentalist faith ("embryos are sacred") that stands in the way of progress. Perhaps the young girl's courage will cause him to question his own rational certainty that the God she worships is simply an illusion, or to see her very desire for God as evidence of God's existence. Yet whatever letter the biologist writes, science cannot tell him what to say. Perhaps it would be better, at times, for the impotent scientist to say nothing.

THE SCIENTIFIC ETHOS

But humility, alas, is not always a prominent scientific virtue, at least among the most prominent scientists, and especially among many modern biologists. Many scientists believe, with compelling evidence, that they are the true benefactors of mankind, the loving saviors of man's broken body, the real guardians of truth. And while science cannot tell us what to do and how to live, it does invite us to embrace a certain view of man's place in the world. Devotion to the scientific method seems to produce an ethos about the meaning of science—a basic belief that we should do whatever can be done, that the pursuit of useful knowledge is man's most elevated purpose, that nothing should stand in the way of man's quest to uncover nature's secrets.

In pursuing this project, experimental science often exhibits a ruthless charity. Modern biomedical research, most especially, often aims to help the weak by using the weak. It aims to rescue the sick, the suffering, and the desperate by using the embryonic, the dying, and the dispossessed. This experimental method blurs the line between those who benefit from scientific improvements and those who might serve as the experimental basis for such improvements. The modern arts of biology blur the boundary between the human subject and the human object. By studying the parts of animals, the organismic whole is lost; the animal becomes simply a biological "model." And by studying human beings as animals, the line between man and the other animals is lost; man becomes simply another biological resource.

This scientific ethos invites, or perhaps requires, an extreme belief in both human greatness and human smallness. The greatness of man is revealed in his conquest of nature: the physical world is a scrapheap of matter upon which the master scientist imposes the utility of form; the human body is a work in progress waiting for the advanced biologist's novel improvements. But science also cannot escape its understanding of human origins—its view of man as emerging from the dust of the ground, without any notion of being created in the image of a perfect maker. In this view, man is both beast and god, rather than image of God.[21] This is why science can conduct the most ghastly experiments on animals (with godlike power) while also worrying about the effect of modern civilization on the animals of the earth. It is also why science can devote so much energy to curing disease while believing that death is nature's way of improving itself.

Most scientists believe that knowledge will advance, technology will improve, human life will get better, if only they are free to do their work, unhampered by irrational taboos. And yet the perfect freedom that science demands is also, in the end, a form of nihilism. Science in itself sets no limits to human action, except those actions that inhibit the activity of science. But since the domain of science is infinite within nature, there is no action that could not, in fact, be redefined as an experiment. Some scientists hold that human beings are hard-wired to behave in certain ways, including ways that are compatible with our bourgeois values. But this faith in human nature in general is not the same as believing that particular human beings have any particular moral obligations. It is a description of how human beings tend to live, not a set of prescriptions for how human beings ought to live. It is a study of man always open to refinement, not an image of man in which all our refinements should be judged.

And this brings us back to Condorcet's haunting legacy: Science destroys the ideal of perfection in the name of endless progress, but its vision of progress ultimately fails to satisfy our deepest longings. In the end, the cost of believing in progress alone seems too high for those seeking a more perfect salvation from the miseries of nature. The hospital chapel may be an architectural afterthought, but the chaplain (not the "devil's chaplain") still usually gets the last word. Our faith in science eventually gives way to our need for faith. We choose the hope of perfection over the ideology of endless progress, but only after trying for as long as possible to believe in both without contradiction.

NATURE AND ETERNITY

Unquestionably, the modern scientific project has been a great success: Our lives would be inferior—indeed horrible—in countless ways without the technological fruits that were always its primary aim. We have gone some way toward correcting the amorality of nature, using nature as our instrument. For the afflicted, scientists and doctors are often the first saviors, restoring normal life when the ugliness of death seeks another victim. The scientist is, sometimes, the personification of love in a method. Yet ultimately, the modern scientific project will always be metaphysically unsatisfying: Its methods do not answer our most permanent questions; its victories are always temporary and its losses always final.

Of course, human beings need to accept the possibility that life itself is metaphysically unsatisfying. The scientific view of nature—as remorseless and indifferent, inhospitable to human longing, a tragic story with no redemption except that which science itself can offer—may be correct. Those who hope for salvation in God need to accept that there is no proof for such a belief, except the present hunger for redemption and the ancient testimony of past generations. And while no doubt there are limits and problems with Darwinian theory, the intelligent design movement seems to have things somewhat backward: they seek proof of God's existence by describing the orderliness of nature rather than seeking hope in God's love because nature is so disordered. Yes, nature is orderly enough to give us hope in the first place, to make it possible to believe that life is more than a cruel joke—a fact captured every time a newborn child seems to know just how to suckle her mother's breast. Nature is miraculous enough to make us believe that it might be the handiwork of a divine hand. But nature is also as twisted as it is remarkable; and if nature satisfied our loftiest moral and spiritual aspirations, the yearning for transcendence might never arise.

At the same time, while many scientists accuse religious believers of zealously imposing their irrational pieties on everyone, some scientists seem to have embraced a new fundamentalism of their own: the belief that Darwinism explains everything important about being human, combined with the passionate need to convert the unconverted and unsave the saved, with tracts like *The God Delusion* and *Breaking the Spell*.[22] Confronted by the aimless nature they so laboriously study, many scientists seem to need a universal, all-encompassing framework to explain their existence. Yet while orthodox Darwinists believe that the law of animal survival explains much of human behavior, they also believe that being a scientist is nobler than being simply a gene-spreading animal. The point of the scientific project is not simply to see ourselves clearly as the beasts we are but to imagine that we possess the cleverness and magnanimity of gods. It seeks not simply to understand the law of death (evolution as we find it) but to wield mastery over life (evolution as we make it). At his heights, the scientist comes to know the experience of bringing order out of chaos; he glimpses the pleasure that God might have felt at Creation, if such a God really existed.

Despite its inherent limits and frequent excesses, there is great dignity in the scientific vocation rightly understood—the dignity of confronting nature's facts in all their beauty and ugliness, and the dignity of seeking to make human life a little less miserable. Science is, or can be, a noble vocation, a realm of human endeavor that invites human excellence, including moral excellence. Against the sin of despair, the scientist stands for action. Against the postmodern revolt against reality, the scientist seeks truth. Thrown into a world that is mysterious, the scientist seeks to bring into light what is so often shrouded in darkness.

The trouble is that most scientists—at least most modern biologists, whose work dominates the public imagination about science—do not seem to reflect much or deeply about the limits of their method, or about the *moral* significance of the ends they seek and the means they use. In the public realm, biologists seem, all too often, like scientific geniuses and moral simpletons, applying rational rigor to their investigations of nature but relying on feeling alone as their only ethical compass. And for all its appreciation of nature's complexity, the scientific mind seems no rival for the Bible or Aristotle or Machiavelli in understanding human complexity. Next to the philosopher, the neuroscientist still looks, all too often, like a fool.

Utopian science is the most foolish, and the least empirically defensible, of all. For despite Condorcet's claims, science is perhaps most necessary precisely because of the permanence of human sin and human evil, not because scientific progress will be the tool of their eradication. We will continue to need vaccine-makers because evil men will make and use biological weapons. We will continue to need missile-defense makers because evil men will make and use ballistic missiles. We will continue to need surveillance-system makers because evil men will always be plotting the destruction of the innocent. Not the inevitable perfection of man in nature but the permanent imperfection of the human soul makes modern science a moral necessity—including, at times, the kind of ruthless experiments that are justifiable only in moments of supreme emergency, when civilization itself lies in the balance.

And no doubt, in the days ahead, there will be many emergencies —anthrax attacks, avian flu, natural disasters, nuclear explosions— when the power of science will serve both the best and worst impulses of man in a world of great darkness. Amid all this turmoil, perhaps our greatest challenge will be trying to recover an understanding of human life and human death that avoids treating existence itself as a supreme emergency, an endless war against nature, a Sisyphean struggle with no Sabbath in time.

Faced with the contingencies of nature and history, perhaps we need to regain the kind of equanimity that faith alone so often inspires. Faced with a world that often seems absurd, perhaps we should not place all our hopes in science alone. The evidence of nature offers no proof that man's redemptive hopes are justified, but also no proof that everything is hopeless. Thinking philosophically about nature does not destroy the possibility of faith in a God who sees nature for what it is: the setting and substance of the human drama, filled with beauty, stained by imperfection, defined by its incompleteness. Nature is filled with the good things it destroys; natural beings yearn for life even as they are born toward death. And one of those natural beings—man— knows that nature, even when mastered by science, will never satisfy our more-than-natural longings. Amid life's many horrors and wonders, that hunger for the eternal will never go away.

CHAPTER TWO
────────────────

The Human Difference

IN THE CONTEST for oddest pronouncement in a State of the Union address, high marks should go to President George W. Bush's call in 2006 for a national ban on "creating human-animal hybrids."[1] Fortunately, the modern biotech laboratory does not yet resemble H. G. Wells's island of Dr. Moreau, that fictional place where an exiled scientist blends man and beast by vivisection. Not even our most skillful, least scrupulous genetic engineers can manufacture humanzees to provide spare parts or serve as semiskilled labor. We are not yet so talented or so depraved.

Yet the President's call to action did not come out of nowhere. If it seemed strange, that is only because we live in genuinely strange times. In China, scientists are creating cloned man-animal embryos using rabbit eggs and human DNA, and in Britain, researchers now have government permission to do the same. In the Caribbean island of St. Kitts, human neural stem cells are being inserted into monkey brains. At the Mayo Clinic in Minnesota, researchers have produced pigs with hybrid pig-human blood cells, demonstrating the possibility of genetic fusion between man and the lower animals.[2]

So far, no one has produced a hybrid embryo using human sperm and animal eggs or animal sperm and human eggs. But this, too, is probably only a matter of time. Many of these man-animal experiments hold out the promise of useful results, like therapies for Parkinson's disease or the possibility of mass-producing designer stem cells. Some elicit a visceral negative reaction, a Levitical sense of a sacred boundary being violated. All of them should leave us wondering: Just how interchangeable is man with the other animals? Could scientists one day recreate one of our extinct, not-quite human ancestors? Or produce creatures, as Wells imagined in 1896, that are "human in shape, and yet human beings with the strangest air about them of some familiar animal"?[3]

Even the National Academy of Sciences (NAS) has been sufficiently concerned to distribute a set of guidelines regarding man-animal chimeras. It recommends that institutional review boards ask the following questions before approving experiments involving the insertion of human embryonic stem cells into non-human animals:

> Might the cell transfer result in the animal's acquiring characteristics that are valued as distinctly human?... If visible human-like characteristics might arise, have all those involved in these experiments, including animal-care staff, been informed and educated about this?[4]

True, the most remarkable possibilities—the creation of monkeys with complex language, or men with working wings—are scientifically unlikely. But, the NAS seems to be saying, what about animals with human organs? Or man-animal hybrids that develop for a few weeks or months in an animal uterus before spontaneously aborting? Sounds like science fiction, perhaps, but better get the review boards ready just in case.

THE IMAGE OF MAN

It should not surprise us that modern biotechnology is raising anew ancient questions about man's standing among the animals. The trouble is, however, that modern biology has also left us bewildered about how to think through such questions, precisely because it has left us bewildered about man himself. At the very moment when our technological cleverness is increasingly enabling us to blur the boundary between man and the other animals, we lack the clarity of the ancients about what sets man apart.

According to Aristotle, man is the only rational animal, the only being capable of deliberately choosing his own path or contemplating the cosmos in which he lives. According to the Hebrew Bible, man is the only being created "in the image of God"; capable of sin, aware of death, with longings for immortality, he is also ruler of the other animals.[5] While the philosophy of Athens and the revelations of Jerusalem differ greatly on what elevates man, they agree that human beings are, or are meant to be, superior to everything else on earth.

Modern science, by contrast, is not so certain. Its radical ambivalence about man is traceable to the mixed marriage that gave it birth. First came Descartes, who treated the whole material world, including the human body, as a collection of aimless stuff to be mastered and manipulated by the human mind, which stands high above it all. Then came Darwin, who assimilated the whole of man, both mind and body, to the rest of biological life in a seamless continuum. Taken together, modern science treats man as both radically similar to *and* radically different from everything else in nature.

Thus, as theoreticians, modern biologists aim to convince us that man is just another animal; as practitioners, they conduct ruthless experiments on other animals for the sake of improving human life. Modern sociobiology declares human pride to be chauvinistic, yet modern biotechnology progresses only through such unapologetic human chauvinism. The scientist's pride in his biological discoveries is humbled only by his belief that pride and shame and everything else are just Darwinian survival mechanisms repackaged in human form.

It is the Darwinian side of science that most worries those who think about man's standing in the universe. While the theory of evolution encompasses many agreements and disagreements, most Darwinians make three claims about the mechanisms of man's *descent*. First, man and the other animals share a primitive, common ancestry. Second, man came into being through a process of genetic mutation and natural selection, with new biological powers emerging inexplicably, persisting genealogically, and accumulating sufficiently to form a new species. Finally, coming into being as he did through the chance unfolding of matter's aimless history, man might never have existed at all. It is therefore quite possible that the universe could have persisted without containing any such rational being, or without becoming truly conscious of itself.

On the evidence provided by nature, Darwin's claim of common descent seems undeniably compelling; man's emergence via genetic mutation and natural selection seems likely; and the possibility of man's never emerging seems all too possible. Yet for all its insights into the development of complex life, the theory of evolution ends before the most interesting questions begin. Where did matter come from in the first place, and with it the latent possibility of man? What is the source of nature's fixed laws, by which the chance process of evolution plays itself out? Why do animals seek to survive and reproduce at all, hungering for life even with its manifold sufferings?

To these questions, modern Darwinian theory has no compelling answer, and its methods are poorly equipped even to initiate the right sort of inquiry. Evolution may explain the mechanisms of man's descent, but not the mystery of his *ascent*, including the wonder he exhibits about the origins and destiny of the cosmos—a wonder that serves no useful animal function. A theory of man's origins is not yet a theory of man, let alone a theory about why there is something rather than nothing.

What is especially striking is the zeal with which contemporary scientists defend the theory of evolution against its skeptics and detractors even as they often fail to acknowledge or understand its limitations. They see polls showing that a high percentage of Americans disbelieve in evolution, and they cringe, feeling, in the words of the Nobel laureate Harold Varmus, "under siege."[6] By contrast, in the conversion of each new child in Pennsylvania or Kansas to a belief in evolution, they see an intellectual and moral victory. For radical neo-Darwinians like Daniel Dennett and Richard Dawkins, evolution is indeed a kind of liberation: proof that God is dead, proof that we are free to make our own gods, proof that we can impose our own moral order on a world governed only by amoral chance. With this absolute license comes, in the scientists' perception, an absolute and exclusive responsibility to ameliorate the physical misery of humankind, heeding the cry of the sick that falls on heaven's deaf ears.

This is how the modern theory of evolution comes to serve the ameliorative ambitions of the modern scientific project, as first envisioned by Bacon and Descartes. For what biology now says about human origins has become the ground for claiming an uninhibited scientific freedom, aimed at correcting the broken life that nature so callously gives us. The Darwinian metaphysic—man as the product of blind chance—becomes a new basis for Baconian science—man as the redeemer of blind nature.

This is why stem-cell research is, along with evolution, the other great scientific issue of the age, where godlike responsibility for human suffering supposedly justifies the godlike destruction of nascent human life. (Here, too, scientists regularly complain of being "under siege.") For many scientists, this divided account of human nature is also the ground for conducting man-animal experiments with virtually no moral limits: because such research will help the sick, and because there is nothing all that special about man in the first place. As the NAS report asserts, "There is general agreement in the scientific community that [species] boundaries are to some extent arbitrary."[7] Mouse, monkey, man—at least under the microscope, it is all a blur.

BEYOND THE QUESTION OF ORIGINS

Given the often sterile, shrunken, and confused image of man put forward by the neo-Darwinists, it is hard not to have some sympathy for the movement known as intelligent design (ID). Against a view of nature as aimless matter in motion, proponents of ID seek to restore a guiding purpose to the universe. Against the belief that human freedom is simply an unintended by-product of blind evolution, they aim to defend human freedom as the Creator's gift.

Describing what motivates him, William Dembski, perhaps the most prominent ID theorist, once declared:

> I think God's glory is being robbed by these naturalistic approaches to biological evolution, creation, the origin of the world, the origin of biological complexity and diversity. When you are attributing the wonders of nature to these mindless material mechanisms, God's glory is getting robbed.[8]

And if God's glory is robbed, man's glory is diminished. The elevation of man that once came from being created "in God's image" is replaced by the will of the robbers, who believe that man is a beast answerable to no god, and that scientists are gods who can remake human life as they see fit.

In essence, ID theory is a modern restatement of the ancient belief that the visible order of the cosmos is evidence of God's design. "The heavens declare the glory of God," the Psalm tells us, "and the firmament proclaims his handiwork."[9] Just as the watch is proof of a watchmaker, the existence of man is proof of a divine engineer who imagined the human whole before breathing life into the parts. Modern ID theorists—like Dembski, Michael Behe, and Stephen Meyer—reformulate the watchmaker metaphor in mathematical and biochemical terms, focusing primarily on the inadequacy of Darwinism as an explanation for the emergence of complex life. Organisms, they say, possess an "irreducible complexity" (the phrase is Behe's[10]), for which the design of the whole necessarily precedes the arrangement of the parts. Even the hidden functions of the body—the parts of the parts, like the molecular mechanism for blood-clotting—make sense only as functional wholes with subordinate parts of their own. Only if every species type came into existence by design, not piecemeal via nature's purposeless mutations, could the organism precede and give direction to the organs, the organs to the cells, the cells to the proteins.

Yet whatever the merits (and limits) of ID as an explanation of human origins, it too offers little as a theory of man's being. Saying that humans are designed says nothing about what they are designed for, or how they are different from the other animals that are also, presumably, products of design. Conversely, to celebrate the orderliness of nature as a reason to believe in divine creation ignores the gross disorderliness of nature that relentlessly afflicts us. Inexplicable natural misery is what often awakens a longing for the divine in the first place, or a desire for perfect justice that will transcend the crookedness of nature with its mad epidemics and childhood cancers. Without a theory of man's fall—an inescapably religious idea—the theory of design seems like a half-truth, if not an absurdity.

The irony is that by focusing relentlessly on man's origin, not man's being, ID theorists ultimately make the same error as orthodox Darwinians. In an age when biotechnology may soon allow us to redraw the biological boundaries between man and the other animals, what we need to understand is not the human beginning but the human difference. Who we *are*, not where we came from, is the question that matters most. For this, we need to look beyond the chimera in the laboratory, beyond neo-Darwinism, and beyond the theory of intelligent design.

A METAPHYSICS OF THE GRAVE

Fortunately, there are other sources of human self-understanding and richer ways to explore the difference of man and the difference it makes.[11] In modern times, perhaps the deepest thinker about man's place in nature was Hans Jonas, a philosopher of biology and a founder of modern bioethics. In a remarkable essay, "Tool, Image, and Grave: On What is Beyond the Animal in Man," Jonas reminds us that our commonality with the animals is hardly news to anyone who has read the ancients. But, he adds, "recognizing these similarities . . . has never been an obstacle to distinguishing" man from beast, or "from perceiving in him something that is beyond mere animality and locating his essential nature in that difference."[12]

Jonas does not dismiss Darwin's discovery. He even welcomes it as a corrective to Cartesian extremism, or the misguided belief that everything in nature except the human mind is formless stuff, to be manipulated by us without restraint. While raising difficult questions about man's stature, Darwin also elevated the rest of nature: the mouse was no longer just a biological model for dissection but an animal precursor of the dissector himself.

And yet, Jonas insists, Darwin's focus on natural selection only takes us so far. Survival value alone "fails to explain the enormous surplus of those characteristics that have emerged in man beyond what is needed for purposes of survival."[13] In this connection, Jonas explores three artifacts—the tool, the image, and the grave—that together reveal the ascending heights of man's difference.

In tools, humans meet their animal necessities with distinctly human intelligence and will. They build hammers and screwdrivers in order to build scaffolding and cranes in order to build bridges and skyscrapers. While other animals sometimes use what they find in nature to achieve some instinctual purpose—Jonas gives the example of sticks and stones used as "momentary aid[s]"[14]—man alone deliberates about his inventions, learns by trial and error, perfects past designs with the latest breakthroughs, and treats nature as a raw material. As our tools become more sophisticated, so do our needs and desires, pointing beyond the utilitarian to the aesthetic. The tools man makes often serve no function except his own love of beauty, and especially his human pride in the aesthetically pleasing artifacts of his imagination.

Indeed, it is in the realm of the image, Jonas's second instance, that man "displays a total, rather than a gradual, divergence" from the other animals.[15] Even the simplest picture creates a new layer of being, entirely separate and separable from the merely material. The decorated cave or painted canvas becomes something other than it is—a reminder of the glorious hunt, a portrait of the Christian savior, a two-dimensional diagram of the three-dimensional brain.

This physical image always begins in the imagination, where man alone can remake past events or bring to life future possibilities. In the imagination, we humans liberate ourselves from the constraints of time and place. No longer enslaved by the iron law of necessity, we assert a novel freedom by remaking the world in our mind's eye—including, most profoundly, the freedom to imagine what lies beyond death.

This brings Jonas to his third artifact. An image can be partially explained by the useful functions it sometimes serves, but "that the grave is an exclusively human phenomenon is empirically demonstrated to us by the fact that no [other] animal buries or gives further consideration to its dead."[16] In the grave, humans confront their self-conscious mortality with rituals of remembrance; they think beyond the lifeless body to the possibility of immortality; and they reach beyond nature in search of an answer to nature's limits. "Metaphysics," says Jonas, "arises from graves."[17] For in knowing that he must die, man is forced to reflect on who he is.

THE SEXUAL DIFFERENCE

Judged by the standards of modern science, Jonas's speculative reflections on man's nature "prove" nothing. Yet by taking seriously these everyday objects and deeds, by looking with self-awareness at the visible evidence of man's ways, he invites us to see what the human animal truly is. Following his example, we might explore still other dimensions of man's difference—other reasons to regard man as within yet beyond the animal kingdom, the rightful ruler and potential despot, whose mastery sometimes elevates him far above and sometimes degrades him far below the mere beasts that are his ancestors.

One of those dimensions, and one that man conspicuously shares with the other animals, is *sex*. Here Darwin himself—the historical Darwin, not the philosophical one—offers an instructive case. Soon after returning to England from his great voyage on HMS *Beagle* in 1836, Darwin sought to decide whether to marry. Being a good rationalist, he prepared two columns of opposing notes, "marry" versus "not marry," each beginning with his thoughts on the meaning of children. Children were a "second life," he observed, but also a threat to his "freedom to go where [he] liked." Children would be companions and caregivers in old age; but they would also mean the "loss of time —cannot read in the evenings" and "less money for books etc.—if many children, forced to gain one's bread."[18]

In the end, even with all the losses carefully calculated—"I never should know French—or see the Continent—or go to America, or go up in a Balloon, or take solitary trip in Wales"—Darwin married and became the father of ten. "Never mind my boy—Cheer up—One cannot live this solitary life, with groggy old age, friendless and cold and childless staring one in one's face, already beginning to wrinkle. Never mind, trust to chance."[19]

When it comes to sex, men and women are animals indeed. He and she desire sexual coupling, usually with a member of the opposite sex; and he-and-she reproduce sexually, the union of bodies permitting the union of egg and sperm, giving life to a new person connected to father and mother genetically but also a wholly unique member of the species. But as one moves beyond this birds-and-bees aspect of things, the sexual behavior of human beings diverges dramatically from that of their animal cousins. Darwin's life, perhaps even more than his theory, points us toward the truth of human sexuality.

For the very fact that Darwin had to decide—that he could deliberately choose *against* children—shows how different he was from the other animals he studied. Only humans bring sex under the kingship of reason, and only humans, seduced by the pleasures of childless freedom, or depressed by the miseries of an inhuman world, or devoted to a calling like the priesthood, can choose against self-perpetuation.

Just as humans alone can choose suicide, humans alone can choose sterility; to humans alone would it ever occur to sever the pleasures of sex from its reproductive consequences and parental obligations. Human beings alone desire, invent, and use birth control, the very name of which—"*the* Pill"—denotes a certain special status among man's other pharmacological achievements.

The sexual difference between man and the other animals obviously does not end there. Both prostitution (the selling of the body, as a thing) and marriage (the giving of the body, in fidelity) are uniquely human possibilities. Moreover, only man can remedy the body's infertility through technological intervention and is capable, potentially, of replacing sexual reproduction with asexual reproduction via cloning.

In this most animal realm, humans reveal how high above the other animals they stand and how far below them they often fall; they reveal their unique dignity, and their unique capacity for self-degradation. Only man could have produced Paris Hilton and Internet pornography, or Jane Austen and the Song of Songs. Interestingly, what is most elevated in human sexuality mirrors the behavior of some animals, who in their own instinctual way give themselves to one another and take responsibility for their own offspring. What is most dehumanizing in human sexuality is distinctly human, disconnected entirely from any animal precursor.

THE FIRE OF MAN

The passion of sex is often described as a kind of fire—sometimes extinguished in a night, sometimes suffocated by marriage, sometimes transformed by hearth and home. But fire, unlike sex, is a natural phenomenon that burns to no end, leaving behind ashes rather than offspring. Unlike man or animals, fire has no aim, no desire, no inner life. It is a power that appears without warning, a relentless agent of destruction.

Unless, that is, it is tamed by men. For animals, fire is neither servant nor symbol. For man, it is both, even while it remains a potential destroyer—a servant of good and evil, a symbol of terror as well as hope. In Aeschylus' famous tale of Prometheus, fire is the source of man's ascent out of "wretchedness," stolen from the god Zeus who sought to keep it for himself.[20] Without the art of fire, a distinctly human life, with cooked food and heated homes, is unimaginable. Yet the many meanings of fire point beyond the utilitarian—from the fires of hell to the Olympic torch to the fireside chat. Armed with fire, humans have committed the most evil acts, systematically burning their fellow humans in ovens, demonstrating man's capacity to be more beastly than the beasts. With fire, the same burned bodies are lovingly mourned in candles of remembrance, tiny flickers that connect the living and the dead with the possibility of the eternal.

Perhaps like nothing else in nature, fire has become the mark of man's difference—the emblem of our creativity (Lincoln's "fire of invention"), our depravity (the fires of the Holocaust), and the mystery of our being (the Bible's burning bush). It seems only fitting that Moses encounters God shrouded in fire, even as the fire of divine liberation is made vividly present in the journey out of Egypt. Our longings are answered, but the ultimate source of our being is left shrouded in mystery.

What best defines the modern spirit, however, may not be the burning bush but the flaming rocket. Recast as conqueror of the heavens, man alone uses fire to venture up into space; and up there, in the vast unknown, he simultaneously displays his greatness and discovers his smallness. "Rocket science" is a metaphor for distinctly human genius, a standard for belittling, or bringing down to earth, the mundane tasks of our animal life. But as Hannah Arendt observed in a 1963 essay, "The Conquest of Space and the Stature of Man," our venture into space also demotes us. "The astronaut," she wrote, "shot into outer space and imprisoned in his instrument-ridden capsule," assimilates himself into the cosmos that he once stood proudly above. To be on a space walk is to confront, in one perilous moment of self-recognition, both man's power and his vulnerability. It is to witness firsthand our creative ascent—into the void.[21]

MAN APART

Of course, sex and fire—or tool, image, and grave—are not the only phenomena of life that invite reflection on the human difference. Eating, like sex, is a phenomenon that we share with the other animals, but only man incorporates food into an elevated way of life. Man alone among the animals dines with old friends in fine restaurants, says grace before meals, and separates milk and meat in observance of the laws of kashrut.[22] Water, like fire, is a blind phenomenon of nature that functions for man as both servant and symbol, and sometimes as destroyer. Water is relief for the thirsty and death for the drowning; it can power cities or destroy them; it is the tranquility of the lake ("beside the still waters") and the violence of the flood ("the end of all flesh is come before me").[23]

In all of these—sex and fire, eating and water, tools, images, and graves—we have a portrait of man far richer than that proffered by "natural selection" or "irreducible complexity," interesting and important as those ideas surely are. In his sexuality, the very function he shares most closely with the other animals, man potentially degrades and dignifies his bodily life. In his exploration of space, the summit of his unique cleverness, man risks demoting himself into a helpless speck among the stars. In his use of fire, an aimless natural force, man reveals his greatness and wretchedness, his creativity and terror, his hope and despair. Beginning with man as he is—the sexual being, the keeper of fire, the maker of graves—we can ascend to the divine possibility that man alone eternally longs for. We can imagine the God behind the burning bush, even if we can never prove His existence scientifically. We can glimpse who we are, if never know whence we came.

Modern science is itself one of the great exemplars of the human difference, the mark of our creativity and, often, our compassion for those afflicted by the natural miseries that science aims to diagnose and ameliorate. But, as we have seen, our science was also born of two radicalisms—the Darwinian reduction of man to the beasts and the Cartesian elevation of man into a god—that occasionally unite to threaten our human dignity. Standing before a new age of man-animal mixing in the laboratory, we need to be mindful of the human animal who is at stake in our experiments: the only rational, moral, God-seeking being. This same being, it behooves us to remember, also stands to be judged for the use he makes of his great powers. It is not just the callous destruction of near-human life that should concern us in an age of hybrids and chimeras. It is also the self-degradation of man, who would lie down with the beasts in his quest to remake nature like a god.

PART TWO

The Ethics of Progress

CHAPTER THREE

Bioethics in Wartime

IN THE MIDST of World War II, C. S. Lewis addressed a class of Oxford students on the question of "Learning in War-Time."[1] With so much uncertainty, so many peers dying in the trenches, so much suffering and destruction, why should any decent person spend his time reading and studying? With so much horror and such high stakes, what room can there be for contemplation, laughter, marriage, and the other pleasures and concerns of everyday life? The answer Lewis gave, in his typical way, was both sharp and deep:

> The war creates no absolutely new situation; it simply aggravates the permanent human situation so that we can no longer ignore it. Human life has always been lived on the edge of a precipice. Human culture has always had to exist under the shadow of something infinitely more important than itself. If men had postponed the search for knowledge and beauty until they were secure, the search would never have begun. We are mistaken when we compare war with "normal life." Life has never been normal.[2]

This insight cuts two ways: We should still laugh, still marry, still light Sabbath candles—if laugh less loudly, marry more urgently, rest less easily. For these are the human things to do, even in a time of war. But just as war should not evict everything beautiful, it does not excuse us from moral self-examination in the midst of self-defense.

At present, America is engaged in a new, uncertain, perilous struggle against radical Islam. Civilized nations face a mortal threat—the "world's most dangerous men" attempting to arm themselves with "the world's most dangerous weapons," as President Bush has put it[3]—and we are right to focus our energies and talents on averting the nightmare of another, perhaps nuclear, September 11. Confronted with such a dark prospect, all other concerns can easily come to seem insignificant.

Which raises the question: Why should we be concerned about bioethics in a time of war? With tens of thousands of American troops in harm's way, with grave threats to life and limb, why should we worry about the fate of embryos so small we can only see them through the microscope, or the prospect of a few renegade scientists attempting to clone a human child, or the misuse of drugs that numb the pain of unwanted memories? Faced with the specter of technological self-destruction, can we really afford to worry about the dangers of technological self-degradation?

BIOLOGY AND THE GOOD LIFE

The first reason for engaging in bioethics during wartime is the simple fact that scientists are still engaging in science, often revolutionary science. This is, of course, mostly a good thing—good for the scientists, and good for the rest of us who benefit from their inventions. Indeed, doing science is part of the American answer to suicidal terrorism: an example of not succumbing to fear and paralysis, an example of our way of life at its best, seeking truth rather than worshipping death. So long as the labs are still standing, the American thirst for new discoveries and new technologies will proceed apace.

But some of this new science, especially the new biology, raises challenges and concerns. It involves new powers over the beginning of human life, and new powers over consciousness itself. Embryo research. Man-animal chimeras. Drugs that alter our understanding of the past and outlook toward the future. Genetic manipulations that transform the process of human aging. Such research requires not just the peer review of fellow scientists, and not just FDA protocols for ensuring safety and efficacy. It requires moral deliberation and democratic debate, and in some cases it requires the willingness to say "no," "stop," or "here but no further." But if we are to set limits, we need arguments for why we need to set limits. We need to be aware of the moral ambiguities, often tragic ambiguities, involved in deciding what to do with our new powers over the body. But we also need moral arguments capable of defending those human goods or values that a limitless biotechnology might unravel or destroy.

The second reason for engaging in bioethics during wartime is the difference between *living for* something and *living with* something. C. S. Lewis, in his lecture, made this point well: "The rescue of drowning men is . . . a duty worth dying for, but not worth living for. It seems to me that all political duties (among which I include military duties) are of this kind. A man may have to die for our country, but no man must, in an exclusive sense, live for his country."[4] Our American soldiers—brave men and women that they are—may continue to die in the war against radical Islam. Or we may die because however well they fight, terrorism will never simply disappear; it may become only more lethal and more widespread. But our soldiers do not live for fighting terror, and we don't live simply to avoid being killed by terrorists. They live with the duty to fight; we live with the burden of being vulnerable to attack; but all of us, as human beings, *live for* something beyond national security. We seek such security precisely so we may live, and live well.

And this question—what do we live for?—is perhaps the central question of bioethics. If biotechnology is driven by the desire to live *better* than we do now, then we need some idea of what a better life actually entails. We need some way to judge whether things have actually improved, especially if improvement means something more than simply living longer or being safer. And we need some idea of what human biology has to do with the good human life.

My best answer to this question—what do we live for?—is that we live for love and we live for excellence. The trouble is knowing what it means to love and what it means to be excellent. This dilemma is often made pointed by new biotechnologies. For example: Is embryo research an act of love, an effort to save the fathers and daughters who are suffering and dying? Or is embryo research a corruption of human love, undermining the respect for life and defense of the vulnerable that are love's highest form and ultimate purpose? Likewise: Are new drugs that improve muscles or memory aids to human excellence— allowing us to achieve more, to do more, to break records, to beat rivals? Or are such drugs corruptions of human excellence—turning us into biological artifacts, as chemically dependent for our achievements as the drug addict is for his transient pleasures? These are—often if not always—very hard questions, but ones we must try to answer as best we can. To answer them requires having some idea of a human life well lived, especially a human life well lived in the age of biotechnology. Bioethics, rightly understood, has an important role to play in articulating this vision—in saying what we *live for*, not just what we must *live with*. We do no service to our soldiers or ourselves if we stay silent on these ultimate questions, or simply say (as bioethics often does) that everyone must decide for himself.

The third reason for engaging in bioethics during wartime is that it is, in a strange way, the best time to do it. A human life well lived is a life lived well in light of our human possibilities, but also our limits; our noblest aspirations, but also our mortality. And war reminds us, like nothing else, that we are mortal. This, too, was seen clearly by Lewis: "The only reason why the cancer at sixty or the paralysis at seventy-five do not bother us is that we forget them. War makes death real to us. . . . All the animal life in us, all schemes of happiness that centered in this world, were always doomed to a final frustration. In ordinary times only a wise man can realize it. Now the stupidest of us knows. We see unmistakably the sort of universe in which we have all along been living, and must come to terms with it. . . . If we thought we were building up a heaven on earth, if we looked for something that would turn the present world from a place of pilgrimage into a permanent city satisfying the soul of man, we are disillusioned, and not a moment too soon."[5]

Let me make this insight more specific: Bioethics is about bodies. And bodies are capable of the most wonderful things—dancing, embracing, thinking, conceiving, laughing, giving birth. But bodies also decline and die. To be biological is to be mortal. My body is mine, but I cannot control its every movement. My body is mine—I am my body—but I do not ultimately decide its beginning or its end, my beginning or my end. And so the things I *live for* cannot be severed from the things I must *live with*—excellence cannot be severed from decline; love cannot be separated from death, including the anguish that comes from losing those we love most. It is this truth about our natures that the grand dreamers of the biotechnology revolution too often forget, precisely because they seek bodies that never decline and souls that never sorrow.

But war (like disease) reminds us that we are mortal; and we are mortal because we are biological. The ultimate aspiration of biotechnology—or the biotechnology project—is to master and use the way our bodies work so that we might live as if we were not really bodies at all, or as if we could always make our bodies do what we want them to do without fail. Bioethics, at its best, reminds us of what it means to be biological—what it means to be born, to grow up, to make love, to have children, to grow old, and to die—always with the threat of having life suddenly taken away from us. In doing so, bioethics aspires to say what it means to live well as biological beings. It tries to connect what we must *live with* (suffering and death) to what we should *live for* (love and excellence).

And this leads me to the fourth (and final) reason why we should engage in bioethics during wartime. In the Cabinet War Rooms in London, the underground headquarters where Churchill directed the war and exhorted the spirits of his countrymen on national radio, there is a poster for sale in the gift shop. It is a giant picture of Churchill's face, with his finger pointing at us, and two words: "Deserve Victory!" If the war against radical Islam is to be a just war, we must not only fight in a just way, but we must fight for a just cause. Our idea of civilization—the American idea of progress and democracy, freedom and prosperity, science and commerce—must be worth fighting for. And one of the greatest tests for whether we will deserve victory in the years ahead will be how we deal with the human questions raised by biotechnology, and whether we can steer our way through three sets of moral dangers.[6]

The first danger is that in the desire to extend life we will become callous towards life, using the weakest among us as tools to keep us strong. This is the moral challenge of embryo research. The second danger is that in the desire to be happy or at least untroubled, we will seek a merely cosmetic consciousness, chemically divorced from the ups and downs of real life. This is the moral challenge of psychopharmacology. The third danger is that in the desire to have a child of "my own," we will develop techniques that confound the very nature of biological origins, or that make normal sexual procreation a second-rate option. This is the challenge of new ways of making babies—including human cloning, preimplantation genetic screening, and the use of gametes derived not from mothers and fathers but from dead embryos and fetuses. Facing these moral challenges is the task of bioethics; and how we proceed will say much about the kind of civilization we are, and whether we are right to plant the seeds of our modern civilization in vineyards (or deserts) around the world.

WHY THIS MOMENT MATTERS

While confronting the ethical challenges of biotechnology involves imagining the future, one might look at modern culture and conclude that the future is already decided. We are already, in important ways, a "eugenic" or "cosmetic" society. We already tolerate or embrace surgical enhancements of our physical appearance, for no other reason than individual desire, and with no limit except our ability to pay. We already advertise, on billboards and in television commercials, drugs like Zoloft and Paxil that promise to make anxious people "happy" and imperfect lives more "perfect." We already pick and choose the genetic characteristics of our offspring—male or female, healthy or sick— taking what we like and discarding what we don't. To argue about new biotechnologies is to pretend that many questions of principle and practice are not already settled. Even as biotechnology advances, American life goes on—neither as terrible nor as wonderful as those in the bioethics debate often claim it is becoming. And for those inter- ested in the state of American civilization, there are many other issues to worry about: the condition of marriage, the frequency of abortion, the state of the university, the abuse of the natural world. To focus on biotechnology is simply the ethicist's effort to capitalize on a scientific novelty, while the real cultural fights—and harder cultural fights—lie elsewhere. There is some truth to this, but it is not the entire truth.

In just the last decade, we have seen a wave of profound biotech- nological advances. In 1997 we cloned the first mammal; in 1998 we isolated human embryonic stem cells; in 2000 we completed the "first draft" of the entire human genome. And in the last few years, we have produced children with three genetic parents; cloned embryos using human DNA and rabbit eggs; and produced human embryos with mixed male-female genders.[7] In the meantime, research pro- ceeds apace on artificial wombs, on single-sex procreation, on drugs to control the contours of the human psyche with greater precision. Much of what was predicted in the 1970s—when the biotechnology revolution began in earnest—seems to be coming to pass, if not always as quickly or dramatically as many promised or feared.

At the same time, the uses of biotechnology for genetic and phar- macological "enhancement" are becoming more apparent, at least as possibilities. This is the subject of a 2003 report by the President's

Council on Bioethics—*Beyond Therapy: Biotechnology and the Pursuit of Happiness*—which is perhaps the single deepest reflection on how biotechnology will change the character of human life in the years and decades ahead.[8] For while the new biotechnology sometimes appeals to the established goals of modern medicine—freedom from "the maladies both of body and mind"—it also points "beyond" them: not only healing ourselves but controlling the genetic make-up of our offspring; not only easing today's suffering but chemically altering our identity; not only treating infertility but remaking the very idea of "being fertile."

Taken together, the triumph of biotechnology "beyond therapy" implies the end of many ethical norms and human realities that we have long taken for granted. Whether biotechnology is the architect of such changes or simply a manifestation of them is hard to say, and ultimately insignificant. What matters is that biotechnology makes these transformations of human life both more plausible and more irreversible. It points to a *post-medical* age, where doctors serve desires rather than treat disease. It points to a *post-sexual* age, where the differences between men and women are no longer essential to reproducing the species, and where modern technique is a "better" way to make babies than having sex. And it points to a *post-personal* age, where individuals can remake their memories and temperaments to become the person(s) they always wished to be but never really were, and perhaps never really are.

We also confront the looming challenge of biotechnology in the hands of illiberal regimes like China. Chinese eugenics and Chinese "medicine"—including mandatory abortions, state regulation of child-rearing, and the harvesting of organs from the living—are by now well known. But in our American optimism about biological and genetic progress, in our belief that the new technology is not dangerous but life-affirming, we have thought little about how our novel techniques will be used by nations with less respect for human life than we now have; or whether the parallel between our science and theirs might suggest that something is amiss in the ethics of our own research.

A few years ago, we heard the tale of an American scientist heading to China to engage in one of the first real genetic engineering experiments, involving the attempt to produce a child with three genetic parents, and ending in the gruesome death of multiple fetuses.[9] Other Chinese experiments—experiments aided by our own basic science—include efforts to implant a female uterus into the abdomen of a man and to produce hybrid man-animal embryos.[10] Can we doubt that as techniques improve for genetically screening offspring, or developing human fetuses inside animal wombs, or chemically altering the memory of prisoners and soldiers, that China will live by the maxim: all things possible?

The conventional American response to such wayward Chinese research is two-fold, and often circular. The first response is that it is not the research that is the problem but how it is used and who uses it; it is not the tools that matter but the ethics of the users. The second response is that you can't stop progress: since countries like China are going to do these things anyway, so should we. And so, while we might pursue certain areas of research for what seem to us good reasons, our capacity to condemn biotechnology's abuses—to make the case for human rights and human dignity against those regimes that ignore them—may one day be compromised if our experiments make us more like them, rather than them more like us. At the very least, it should strike us as strange that an authoritarian regime like China is seen by some American researchers as a haven for "free" science.

OUR CULTURAL DIVIDES

Perhaps the reason that bioethics issues so often grip the nation is that they aggravate the most fundamental divides within American culture, especially between secular liberals and orthodox believers. For while there are many non-religious reasons to be concerned about where biotechnology might be taking us, the truth is that most opposition to the "Brave New World" comes from people of faith, and mostly Christians. Such people make rational arguments in the public square

—often more rational than secular scientists, who frequently defend the morality of their experiments with largely emotional appeals—but they are moved to argue by their belief in God and His commandments, especially regarding the sanctity of human life and the dignity of human procreation.

In aggravating these political and cultural divides, the biotechnology project cuts to the very foundation of the American project. It challenges our idea of equality, and raises the existential problem of inequality in new ways. Why do some people get sick and other people don't? Why are some born with the capacity to be Olympic athletes and others aren't? Those who oppose biological enhancement often argue, I think rightly, that the dignity of an Olympic runner would be lessened if that person were genetically or biologically enhanced. But one has to wonder about the meaning of appealing to human *excellence* to make the case against human *enhancement*; of looking to the best among us—the most gifted and most talented— to argue that we should not make ourselves artificially better. Most people—no matter how hard they work—will never have the genetic equipment to be Olympic runners. These natural inequalities, which seem inherent to being human, are even more pointed when it's a matter not simply of *mediocrity* but *disease*; when one thinks of sick or dying children; when one comes to believe (understandably, but perhaps falsely) that such diseases are an *injustice* that we have an obligation to correct by any means possible.

And here we see the divide between secular liberals and orthodox believers most clearly. Part of the reason many people of faith (mostly Christian) are so willing to oppose aborting "imperfect fetuses" or destroying embryos in the pursuit of health is that their faith provides good answers to the problems of human limitation, suffering, and mortality. Biblical religion teaches that there is an inherent dignity that comes with creation, a dignity that all human beings possess at all stages of life, simply by virtue of being one of God's creatures. And while people may suffer in this life—with disability, disease, imperfection, and death—they can be saved in the next one. Their wretchedness is a pilgrimage. Modern liberalism, by contrast, has a more difficult problem trying to explain why people are born with great natural inequalities; or why, later in life, we'll all be unequal to the vigorous selves we once were. Without such answers, the temptation to become liberal eugenicists or libertarian seekers of eternal youth is very great.

The Declaration of Independence says that "all men are created equal," but in some unfortunate ways this is clearly not so; nature's inequalities are often harsh and always blind. And in some ways the American regime, on its own terms, doesn't have a satisfying answer to the unequal afflictions of the body. There is no American book of Job except the book of Job itself. Maybe biotechnology is our modern answer or solution. Maybe we're going to make ourselves biologically more equal—especially more equal to pursue happiness. This seems to be the guiding sentiment of liberal humanitarians, both scientists and politicians, who defend research on embryonic stem cells. They want children born with grave diseases to live full lives—like everybody else. They want fairness where fate or genes or both has denied it. They revolt as Job did—"The Lord denies me justice!"—but they do not look where Job looked for an answer.[11]

FINITUDE AND EXCELLENCE

The question is what are we willing to do in the pursuit of such justice, and to what extent can we coerce the world—the world of nations and the world of nature—to become "right." Perhaps it is a fitting accident of history that the bioethics issues and the war on terrorism both came to dominate American public life at the same time—with President Bush's stem cell speech delivered on August 9, 2001 and the horrible attacks following a month later on September 11. Both of these challenges have demanded a new moral and political seriousness; both raise questions about life and death, about the prospects for human happiness in modern democracies, and especially about our idea of the good life and good society. The bioethics debate—and biotechnology itself—is driven by our dreams of greater health and better lives; by our quest for greater control over the course of life from birth to death; and by the belief that our biotechnical ingenuity might eventually conquer the worst human diseases. And yet, the war on terror reminds us of the mortal fragility of life, and the dark side of modern science, and the permanent burdens of history that biotechnology alone will not likely conquer. As Lewis reminded us: "The war creates no absolutely new situation; it simply aggravates the permanent human situation so that we can no longer ignore it."[12]

But to be awakened to the permanent dangers and imperfections of life is also—hopefully—to be reminded of the good things that are endangered. And this brings us back from the current biotechnology moment to the deepest aspirations of bioethics: to connect what we must *live with* to what we should *live for*. To be biological is not simply —or most importantly—to be on-the-way to death. It is to be alive, and to be alive in a fittingly human way—embracing a spouse, running a race, laughing at a joke, singing a song, reading to a child. The philosopher of biology Hans Jonas understood this perhaps more deeply than anyone: "The fear of death with which the hazard of this existence is charged is a never-ending comment on the audacity of the original venture upon which substance embarked in turning organic."[13] That is to say, the audacity of striving, aspiring, self-conscious human life, of beings who love in the face of death and seek perfection even in the face of inevitable decline.

To do bioethics well, we need to understand what human biology has to do with a good human life, and what the fact of our being *finite* has to do with the possibility of our being *excellent*. War is often the great arena of both these elements of our natures—our finitude and misery, our excellence and greatness. But it is not the only arena.

To understand the excellence of human beings—biological human beings—one must consider the double meaning of the fact that such excellence might not have happened at all: because the beings capable of being excellent might not have *been* at all; and because the beings capable of being excellent might not have *performed* excellently. There may never have been a Tiger Woods (with his body, proportions, coordination, desire), and Tiger Woods might never have cultivated his given (biological) possibilities. He might rise to the occasion or might be overtaken by it, and he might come back again next year or he might never return to his lost summit. The first insight points us toward the *givenness* of human excellence; it awakens gratitude for the beings we are, or resentment for the beings we are not. The second insight points us toward the *madeness* of human excellence; it awakens aspiration, nobility, and pride in one's own agency and achievements, or despair in the face of one's own failures. Both these intuitions stand in opposition to the spirit of the bioengineer, who seeks an excellence that is broken down to its parts (by himself; for his patients) and guaranteed to work (like a machine) every time.

To be biological is to live this double truth or two-sidedness of our nature: It is to experience one's own accomplishments as one's own doing. It is to be what one does in the very act of one's own doing it. It is to be, in the deepest sense, a *self-made* man. But to be biological and to be human is also to realize the limits of becoming what one wills oneself to be by one's own efforts. It is to will the healthy body even as it fails, or to will the graceful movement even as one is clumsy. We can imagine it—health, grace. We can imagine ourselves as having it—being healthy, being graceful. But we do not have it, and we cannot always (or ultimately) get it. The body fails.

On one level, this points to the unity of mind-and-body—to the fact that we are our bodies; our bodies are not simply tools for our use. It is not simply that the body fails us. It is that we, as bodies, fail. But on another level, it points to the experience of alienation inherent in being biological: realizing that our heart beats without our making it beat; being filled with gratitude for our own given biology; then realizing that just as the heart beats without our doing, it can (and will) stop without our doing. Our biology (our life) is both given and taken away. To experience this alienation is to look upon our bodies as if looking upon someone else or something else. This detachment is the basis for modern biotechnology (mind working on body to shape body to mind's will) and the basis of philosophical wonder and religious awe: We realize that our bodies are both ours and not ours, and that they do not obey our every command. And our capacity to realize this fact means that we both are and are not simply our bodies. We have the distance of self-regarding self-awareness, at the cost of not fully "living immediately," like the animals do—but with the gain of aspiring to more than animals ever can. We will decline and disappear, and we know it. But we can also hope for something beyond what we are now. We can hope for life in another form—whether as post-biological minds without bodies or post-biological souls resting with God. Or we can simply take pleasure in the fact that we can imagine, with our own reason, the very types of perfection that we can never embody because we are bodies. And we can do our best to imitate such perfection, to live well knowing that we will never become perfect ourselves, to take pleasure in the rare animal-like moments when we can and do live immediately.

To be biological is to experience the alienation of one's own desire to be more than biological—or even the desire to be perfectly biological, a desire that is ultimately paradoxical. It is to witness ourselves (our bodies) becoming something (declining into something) we wish they would not, and learning to accept that our agency is real but limited. The very things we do (or once did) as ourselves were always given possibilities and lost possibilities, once given and to-be-lost. But they are no less real and no less excellent for being finite. The possibility of decline also suggests the possibility of a summit from which we fall.

A true bioethics—on the precipice between life and death, finitude and excellence—wrestles truthfully with the meaning of our biological humanity. It seeks to explain what our nobility and our decline, our greatest doings and ultimate limits, our body gracefully at work and our body gracelessly not working, have to do with one another. One should admire the fallen soldiers of the world who fight back from death's brink with willful ferocity. And one should admire the wise old man, or the wise young man, who sits or lays down, waiting for death, neither seeking it nor seeking desperately to turn it away, neither so confident in what is coming nor so afraid that he seeks some blinding chemical remedy. This is, it seems to me, the kind of excellence—*living with* and *living for*—that bioethics should defend, and biotechnology can never produce. War reminds us of these existential truths, but they define who we are even when the guns fall silent: our limits and possibilities are written into our flesh, and the efforts to rewrite body and psyche will only fulfill their grandest ambitions by deforming what is best about our humanity.

CHAPTER FOUR

The Embryo Question

FOR THE PAST SEVERAL YEARS, America has engaged in an intense debate about whether to destroy human embryos in the quest to cure certain deadly and debilitating diseases. In confronting the embryo, American civilization itself is under the microscope. We are invited—indeed required—to reflect anew about the meaning of human equality, the sting of human suffering, the nature of human origins, and the spirit of human progress. Few political issues are so fundamental; few moral questions are so complicated and profound.

Our moral ambivalence about human embryos actually goes back to the beginning—that is, to the creation of the first embryos outside the womb. Consider the following two remarks, both made by Robert Edwards, one of the pioneers of in vitro fertilization (IVF). The first was his reaction to the birth of Louise Brown (the first IVF baby) in 1978: "The last time I saw her," he said, "she was just eight cells in a test-tube. She was beautiful then, and she is still beautiful now!"[1] In other words, he was awestruck at the continuity of the person he knew "then" and the person he knows "now"; at the continuity of her beauty, her presence, and her being.

The second passage came only a few years later, in his 1980 memoir about the birth of IVF. "Will we," Edwards asked, "be able to extract the stem cells of various organs from the embryo, the precious foundation cells of all the body's organs and then use them therapeutically? Will it ever be possible to use the cells to correct deficiencies in other human beings—to replace one deficient tissue with another that functions normally? For instance, will we be able to use the blood-forming cells of an embryo to re-colonize defective blood-forming tissue in an adult or child? And will these notions be met with pursed lips and frowning faces?"[2] Dr. Edwards, we should be clear, was endorsing this prospect of destroying embryos for stem cells, and he saw such "pursed lips" as a kind of backwardness, indeed as a religious rejection of progress itself.

And so in only two years, this beautiful being—the embryo he knew as Louise Brown—had become a potential resource for us, there to be exploited and destroyed. The embryo's meaning now depended entirely on its destiny—a destiny held entirely in our hands. What was once beautiful was now merely useful—if useful in the humanitarian project of easing suffering and combating disease. And now that Edwards's stem cell thought experiment of three decades ago is much closer to reality, the stakes of his moral confusion are very great indeed.

To sort out this confusion, we need to recover a richer sense of how the embryo question emerged in the first place. It is easy to forget that this encounter with the early-stage embryo is a novel experience in human history, and that the reason we initiated human life in the laboratory was to answer the longing for a child in the face of infertility. The yearning for a child of one's own, flesh of one's flesh, is a primordial desire. The frustrations of barrenness have long been a source of misery and anguish, and a reason to call upon the aid of Providence. Every civilization has its tales of infertility overcome by divinity, exemplified in the Hebrew Bible in God's answer to Sarah's bitter laughter with the miracle of the first child of the covenant.[3] The technological answer and the divine answer to the pathos of infertility are surely different in meaning. But we all recognize the powerful desire for a child, and understand how the drive to give birth to our own replacements moves us to initiate life in the laboratory when nature does not allow life to begin in the womb.

The second origin of the embryo question is the relentless drive of human curiosity—especially the desire to uncover how we came to be, to understand our own human beginnings. In this sense, bringing the embryo outside the body is the quintessential scientific act of bringing what is in darkness into light—of bringing the embryo from the darkness of the womb to the light of the laboratory. In acting upon such curiosity, the scientist always lives daringly on the narrow ridge between the sacred and the profane—between seeing reality as we imagine God sees it, and usurping those boundaries of reverence that preserve our humanity as neither beasts nor gods. Not everything, after all, is meant to be seen by everybody, or anybody. The preservation of the sacred—especially when it comes to the nakedness of the flesh— requires a certain modesty, a certain reluctance, even a certain shame.

And yet, the reason science seeks to uncover darkness is often not curiosity alone, but the humanitarian desire to ease suffering and cure disease, to answer mortality through medicine. Throughout history, we have shattered many old taboos in the name of healing—such as transplanting organs from the dead into the living—and mostly for the better. But not *always* for the better. And so the question before us now is whether the shattering of yet two more taboos—by initiating embryonic life in the laboratory as an answer to infertility, and by destroying embryonic life in the laboratory as an answer to mortality—is progress, regress, or both at once.

IN THE BEGINNING

In addressing this question, it makes sense to begin with the beginning—with the human embryo in the first few days of existence. The early-stage embryo is a mystery—and a profound one. It is unambiguously a stage in the unfolding existence of an individual or, more precisely, it isan individual at the earliest stage of his or her biological life. We all began as embryos, destined then to become who we are now, with no clear biological markers or quantum leaps to break this continuity of development.

And yet, the early embryo is clearly different—certainly our encounter with it is different—from human beings at all other stages of life, including fetal. The early embryo can easily be mistaken by the human eye for something other than it is—a mere cell, an animal embryo, or a "parthenote" that functions like an embryo for a short time but then dies. We can only see the early embryo under the microscope, and we can only know its presence in the early days of its existence because we created it outside the womb—the womb where it naturally arrives without our knowing it at all. These first days are the only stage of human life where the normal human form is not yet manifest in any way. And even when we know what the early embryo "really" is and why it is really "one of us"—even when we have mastered the biology of embryological development in detail, and thus know the continuity of human life through all its stages—believing what we know is not always easy, especially when those we love might benefit from heeding our untutored eyes, or from giving in to the promises of our most talented scientists.

I can already anticipate the voices of Catholic theologians and pro-life intellectuals, warning me about the dangers of abandoning moral reason to the prejudices of sight or the sentiments of the age, and reminding me how reason corrected our false belief that slaves were not human, or that the black man before my eyes was not a man. The eyes blur reason, they would say. But the eyes can be fixed once we reason correctly. The point is an important one, and well taken.

But one must also remember the following: Reason alone did not teach us that slaves were men; the question of slavery was settled, finally, by a bloody war that vindicated Lincoln's rational arguments. And we must remember that the humanity of the slave was far more obvious—the eyes and sentiments were much more truthful guides, more easily corrected by reason—than the humanity of the early embryo, especially the early embryo outside the body. Conversely, sometimes our eyes and our sentiments are better guides than our theories. Indeed, Peter Singer's defense of infanticide is perfectly logical, though horrible.[4] There may be such a thing as right reason, but there is also such a thing as wrong reason rationally defended. And there may be choices where reason alone is an insufficient guide for knowing what to do, or moving us to do what we know to be right.

And so: Thinking about the "embryo question" is a very strange business—and very complicated. One becomes frustrated with dishonest (or morally obtuse) scientists who describe the embryo as nothing special even as they desperately try to get their hands on it. One becomes restless with pro-life rationalists, who reason as if they can "prove" the equal humanity of the early embryo—a restlessness that should be tempered by a deeper appreciation for their rigorous arguments, and a recognition that their arguments are true, if only part of the story. One occasionally thinks it is absurd to obsess about embryos in the middle of "World War IV" or in the capital city of the world's most powerful nation; or peculiar to split hairs about the fine differences between zygotes and clonotes and parthenotes, believing that the moral fate of the nation hinges on using only this one and not that one. And one worries that our passionate concern about embryo destruction might even distract us from other, perhaps deeper, corruptions of human dignity at the hands of biotechnology—such as the dulling of human aspiration through new psychotropic drugs, or the corruption of human procreation through new ways of making babies.

But upon sober reflection, it is clear that much about the American character is at stake in the embryo debate, as well as much about the limits of reason, the tragic nature of politics, and the moral prospects of modernity. Those who care most about embryos care also about much else that is humanly good and deeply imperiled. And many of the deepest conversations in American public life over the last few years have begun, shall we say, at the embryonic stage—only to develop far beyond it.

So this is where we are: seeking wisdom about the smallest human beings, which set before us the biggest human questions. Four sets of concerns seem most important.

First: How does the embryo debate fit in with bioethics as a whole, and with the moral concerns many people have about the new biology and new genetics? When conservatives, in particular, see trouble ahead with "where biotechnology might be taking us," what is it that they are trying to defend, and do all the things they are trying to defend cohere?

Second: How does a mighty, technological, democratic civilization think about its obligations toward the tiniest human organisms? How does our role in the world relate to our debate about the ethics of biotechnology at home? How does a belief in American greatness (or the virtues of the American way of life) fit together with fears about heading toward a Brave New World? And how does our promotion of democratic progress around the world relate to our concerns about the morality of progress at home?

Third: How do we make sense of our encounter with the early-stage human embryo—when sight and sentiment are perhaps an unreliable guide to what (or whom) lies before us? Is it "rational" to "love" an embryo? Is the early embryo different from the fetus, or the infant, or the adult, and if so, how? Alternatively, if reason can demonstrate why "personhood" begins at conception, is it enough to guide us in the moment of decision, when we may be called upon to accept death in order to "choose life," or to accept suffering and mortality rather than profit from the creation and destruction of the embryos that might save us, or our child?

Finally: What would a prudent politics of the embryo look like— given the kind of nation that we are, the moment that we live in, and the truth about the embryos themselves and our human encounter with them? How do we disentangle the three separate issues of abortion, embryo research, and new ways of making babies? And is there any chance of reaching some tolerable and principled compromise in this debate—one that prevents the worst horrors, halts a deepening moral divide in the nation, and stops us from becoming the kind of people that believe nothing is horrible if it might make us healthy?

These are the kinds of questions we must ask if we are to understand the relationship between embryo research and the American character, if we are to engage in a richer bioethics that is both philosophically deep and politically serious, and if we are to make sense of the prospects for American civilization, at home and in the world.

LOVE AND EXCELLENCE

The debate about embryonic stem cells typically centers on some version of the following question: "What is the moral status of the human embryo?" Is it a person, a mere cell, or something in between? Can we use embryos for research or must we give them equal protection before the law? Can we use them only as a "last resort" and "with tears"? Or can we use them boldly and without remorse? What we seem to be asking when we ask about moral status, in other words, is whether embryos should or should not be inviolable before the law; and whether embryos are or are not members of the human species, and therefore deserving of the same respect, and rights, and neighborly love that democratic societies grant to everyone who is, in official bioethics parlance, a "person."

Now, there are good reasons—good democratic reasons—to ask these kinds of questions. Democracies, after all, set a fairly low bar for granting moral status—much lower than most societies in history—and it is generally a good thing that they do so. It is central to our belief in natural rights, our commitment to tolerance, and our founding idea that all men are created equal in the ways that matter most.

And yet, the word "status," by its nature, is a word that demands degree. It is an undemocratic word, a hierarchical word. And the word "moral" is a word that means more—or should mean more—than simply whether someone is inviolable before the law or a member of the human species. To those who wish to use embryos with impunity, the language of "moral status" is a fitting rhetorical weapon—allowing them to attach status only to those characteristics that embryos lack. And to those who believe that what matters most about embryos is whether they are members of the human species, the term moral status does not give offense, because they believe sound moral reasoning can win embryos and everyone a 100 percent score on the status scale.

But the current discussion about moral status misses much that is most essential, and most interesting, and most puzzling. While it is true that many essential things about human life are held in common by all individuals, there is much about being human that is not held in common at all. And there is much about being "moral"—or having "moral status"—that goes beyond simply whether an individual has rights or is a member of the species. As Harvey Mansfield once aptly put it, the belief that all men are created equal—or that all men *are* equal—is the "self-evident half-truth" of the American Founding.[5] Some men, in other words, are better than others—more excellent, more beautiful, more noble, more angelic. And it is a good thing that some men are better than others. They are our teachers, our models, and our guides. They are the ones we admire, and honor, and sing about, and write about. Whether the greatness of the great is more a gift than an achievement is a difficult question—one central to considering the uses of biotechnology for so-called "enhancement." And while some forms of excellence are obvious if rare—Mozart and Shakespeare come to mind—others are more mysterious, less easy to see, perhaps even more lasting for being so counterintuitive.

At the same time, the half-truth of equality is self-evident in the opposite direction: some men are clearly *not* born equal in ways that matter a great deal in a society that prizes the pursuit of happiness. These individuals are born sick, or disabled, or limited in ways that an egalitarian democracy sees as unfair and intolerable. And so the sick child becomes the great problem—the great injustice—that liberal society must remedy. And liberals seek such remedies in the strangest, often the most illiberal ways: by screening and aborting "imperfects" before they are born; by pretending that the disabled are simply "differently-abled" and thus "equally-abled"; and by defining "personhood" by the higher functions, so that we might use embryos with impunity to heal those who are losing the higher functions.

And here, perhaps, the two major concerns of a conservative bioethics—the greatness of the great and the dignity of the weak—come into focus. Conservatives admire the great humanity of those who run and swim and compose and fight, and they fear a Brave New World where the aspiration to excellence is smothered by pharmacological contentment, or where excellence becomes more artificial than real, more machine than human, or so technical that only the technicians can understand it. But conservatives also defend the dignity of those who will never run or swim or compose or fight, and the dignity of those embryos that cannot yet do these things. And they argue against those who claim that the very lack of these powers makes such lives not worth living or protecting, and against those who are tempted to seek equality by aborting (or euthanizing) the imperfects.

In other words, conservatives are for the highest human types and the most vulnerable human types. They are for unconditional love and conditional excellence. They are for treating seemingly unequal beings (like early-stage embryos) more equally, and for treating truly unequal beings (like Olympic athletes) less equally. They are against screening and aborting individuals with low IQs, and against treating individuals with low IQs as valedictorians—or drugging them so they have the self-esteem of valedictorians.

In the end, the Brave New World should frighten and disgust us because it is a world without love and a world without excellence. It is a world where nobody aspires to anything lofty, noble, or daring, and where nobody must love another when such love is fragile, mysterious, and hard. Conservatives accept mortality and believe in greatness—including the greatness of those who accept their own mortality with great dignity, or who hold the hand of the dying while they die, and perhaps in the final nakedness of the other see their own mortal fate.

HUMILITY AND MIGHT

And this brings us to my second question—the relationship between being a nation that destroys nascent human life (in the quest to remedy the problems of nature) and a nation that exerts force in the world (in the effort to combat the enemies of civilization).

On a day-to-day level, it is surely a good thing that policymakers and generals are not devoting their intellectual energies to reflecting on these two dimensions of American civilization. But at some point, our moral life requires that we ask: What do our anxieties about the morality of progress at home and our desire to spread the culture of progress abroad have to do with one another? What do fears about the Brave New World, or Slouching Towards Gomorrah, or the End of Democracy, or the Culture of Death have to do with our promotion of American values around the world, and our confidence that America's values are good? The simple answer to these questions, of course, is that while America is imperfect—and does not always live up to its highest principles—the *idea* of America is noble and decent. There is some truth to this simple explanation, but it does not explain everything, and it does not probe far enough.

For the fact is, conservatives have always been wisely ambivalent about "progress," and fully aware of both the permanent problems of the human condition and the leveling of religious awe and human excellence that liberal democracy incurs. The same neoconservatives who want to promote modern democracy in the Middle East have perhaps the deepest sense of what is wrong with modern democracy at home, and with modernity as a whole. The insight of neoconservatism is to combine tragic realism and American optimism; to combine a sober awareness of the world's dangers and an unabashed confidence in the American future; to combine a deep sense of just how fallen, and indecent, and decadent we are, with a politics that attempts to redeem, and spread, and ennoble the American way of life. Neoconservatives recognize that the central political problem of the age is spreading the fruits of technology, progress, and liberty abroad, while reining in the excesses of technology, progress, and libertinism at home. They seek progress without expecting our advances to be permanent. And they seek to avoid both the misguided expectation of human perfection through modern technology and the paralyzing despair that high-tech radicalism will destroy us.

But even this does not go far enough. For to think about microscopic embryos and mighty nations is to think about the smallest and the largest human things. It is to think about the morality of American power—over nature, human nature, and human life as a whole. It is to think about the need to nourish the confidence that never surrenders to evil, and the need to surrender to death when opposing it would require transgressing the very moral principles we hold dear. "Surrender" and "Never Surrender" are not bad conservative slogans, each taken in their proper place, and each seen in a properly tragic light.

Without making too much of it, let me suggest two things: First, that to love the embryo (the smallest of human things) is perhaps, in a strange way, to redeem the imperfect, worldly, lethal work of defending our imperfect civilization against its most barbarous enemies; it is to embody an ethic of love that those who live in the world of force cannot usually live by. Second, let me suggest that supporting just wars for democracy using surgical force, and abstaining from the surgical destruction of embryos to heal those whom we see as "unjustly" ill, is not finally (or simply) a contradiction. Both dispositions embody a courageous realism about human evil and human limits, and about the need to face evil and accept limits without becoming evil ourselves, and without believing that we can perfect human life with either sword or science. We don't carpet-bomb cities for the same reason we shouldn't harvest fetuses for research—even if doing so might improve our own health and safety.

The debate about embryos is the right debate for a mighty nation to have. It is the right test of our character, and the right reminder of the proper limits of our dominion in the world, our dominion over nature, and our dominion over human life as a whole. There is much more here to say, but this much will suffice for now: Courage, the first of the virtues, is the virtue most needed on the battlefield and in the sickbed. The alternative is a *decadent weakness,* like the United Nations, and a *decadent strength,* such as a society that uses the seeds of the next generation to profit its own.

THE PROBLEMS OF SIGHT

This leads us back to the mystery of the early-stage embryo—what it is, what to think about it, and what to do about it. Calls for "courageous realism" are all well and good, but we should not fall prey to the reverse temptation of glorifying tragedy, or death, or martyrdom. If there is no sin or error or indecency in using early-stage embryos to save the sick and help the suffering, then we should want to use them. We would be monsters not to do so.

But what are these beings, we must first ask, that we would be using? And who should decide whether to use them, and on what authority? Should the sick decide—since it is they who know firsthand the horrors of disease, who suffer daily on-the-way to death and with the dimming hope for cure? Should the healthy decide, since it is only their distance from suffering—and perhaps their unclouded reason —that might allow us to set limits on medically promising, but morally compromising, research? Should the scientists decide—since they know best of all which areas of research have the greatest chance of success? Or has the scientific vision clouded our understanding of what the embryo really is, and what is truly at stake in our using it?

The first problem we encounter when trying to understand the early-stage human embryo is the strangeness of reasoning about this being at all. This is not to say that we should not reason about the human embryo, or make moral arguments about it, or seek to understand the underlying biology of embryological development more fully and precisely. We surely should, and we surely can, and we surely must. It is simply to note that we are reasoning about something that is by nature mysterious, something conceived naturally in darkness, something whose presence in cases of natural conception cannot be known until after its unannounced arrival. Before I V F allowed us to create human embryos outside a woman's womb, we never encountered the earliest-stage human embryo when it actually existed; we never knew it was there when it was actually there; we only traced its shrouded presence looking back, once we came to know that a pregnancy had begun, and once the developing life was more fully formed.

Some of our earliest thought about human embryos—before modern biology—attempted to give an account of these mysterious beginnings—not to lift the veil, but to understand something of what was hidden underneath. Call it the "first problem of sight," or the problem of needing to understand something significant that we could not see with our own eyes or examine with our own hands. We knew something was there and something was happening—that a child was developing in the nine months between the cessation of a woman's cycle and the birth of a crying infant. But we had no way of directly studying the "process" by which this development took place. And yet, we needed some way of knowing how to treat this being that we could not know firsthand. We needed some way of knowing how to regard "lost seed," or how to regard accidental and deliberate abortion. We needed to know, in these circumstances, whether to mourn or not, and whether to punish or not. Despite deep disagreements in ancient and medieval thought, there seems to have been a widespread intuition: namely, that there is a stage of development after conception but before the "human form" has taken shape, just as there is a stage of development after the human form has taken shape but before birth.

With the coming of modern biology, we came to understand the stages and nature of human development in the womb, and eventually were able to take pictures of this development in process. For the first time, we could see what we had not seen before—the early life-in-process, moving from one stage to the next, from unrecognizable human beginnings to recognizably human form. But even today, we cannot see the embryo at the very earliest stages inside the natural womb; and we cannot know that a pregnancy has begun until at least

a week or so after conception. For all the light we have shed on embryological development, the early embryo conceived in the womb still
arrives as a mystery—not known and not seen at the moment of its
creation or the first few days that follow.

With IVF, we created human embryos outside the body—by unit-
ing sperm and egg in the laboratory—bringing the very earliest stages
of embryological development to new light. The significance of doing
so is something we have barely begun to fathom; it is a boundary we
crossed with little forethought and little reflection; and it may turn out
to be a profound turning point in the history of human life and human
culture. All the absurdity of the embryo debate, and many of the dilemmas of our new reproductive technologies, stem from this new reality.

For the first time, the human embryo was present to us from the
beginning; a life-in-process in the laboratory, but with a limited life-
span, unless its makers "return" the orphaned embryo to its natural
setting, or until we develop artificial wombs. As a result, we now
encounter the "second problem of sight": we can see the early embryo
clearly before us, at least with a microscope, and yet its meaning seems
inadequately explained by what we see; its significance is obscured by
the very act of looking upon it. Is it a mere cell or an individual life-in-
process? Is it a human embryo or an animal embryo? The untrained
eye cannot know simply by seeing, and the scientifically trained eye
comes to see such differences in a way that erodes their significance,
and to see such organisms simply as potential resources for our use.

This problem is unique to the early-stage embryo, and it begins to
reveal why the human encounter with the early-stage embryo is a
unique human relationship, unlike all the others. Before us indis-
putably stands a human life-in-process. And yet, who can deny the
difference in our moral reaction in seeing a fully formed fetus or newborn baby dismantled for its parts, as opposed to seeing an early-stage
embryo disaggregated for its stem cells? This is not to say that the acts
are morally different. They may be or they may not be. But it is to say
that we encounter them differently; and that tolerating the deliberate
destruction of fetuses for research may have a profoundly different

effect on the character of a society, or the soul of those doing the destruction, than tolerating the destruction of early-stage embryos. In the one case, we must forcibly weaken or overcome a natural revulsion that is already there. In the other case, we must awaken a revulsion that is not naturally present.

And so we turn to reason for help: What guidance, we ask, can reason give to sight and sentiment? How are we to reason rightly about the human embryo, especially the early-stage embryo outside the human body, so severed as it is from its natural human context? As Yuval Levin describes:

> We look at this creature, which has been manufactured, molded, formed, examined, and up to a certain point developed under the lights of the laboratory. It is growing, but can only grow so far without further biotechnical intervention. It is living, but only because the scientists have created it artificially. It is human, to the extent that our humanity is in our genes and our potential. It is useful as a resource for medical research, but would develop into a mature human adult if implanted into the body of a woman and permitted to grow. What in the world are we supposed to do with this thing? How is ethics supposed to serve us in this circumstance?[6]

We all know the stakes, as Levin suggests. Having created embryos outside the body, we have discovered uses for them that have nothing to do with what they are by nature—human beings at the earliest stages of a developing life. Instead, we wish to redirect the special powers of these unique organisms to very different ends—to ends having to do not with the promise of "natality" but the miseries of "mortality," not with the continuation of a new life but the hope of saving a much older one.

And this leads us to the problem of reasoning about the early-stage embryo—of rationally discussing a being that is by nature mysterious. It leads us to see the fundamentally different ways we might reason about embryos: giving an account of "what" they are or giving an account of "how" they work; giving a biological account of their continuity with other stages of human life or a biological account of all the things embryos cannot do or don't possess in comparison with most other human beings. Both those who seek to defend the inviolability of human embryos and those who seek to use embryos for research typically proceed by inspecting and dissecting the embryo's properties: in the one case, to prove the embryo's humanity by demonstrating the continuity of biological development from conception to birth and beyond; in the other case, to prove the worth of conducting further embryo experiments, by demonstrating the unique power of embryonic cells to help post-embryonic human beings.

In the end, I believe the pro-life rationalists have the better argument, at least from the standpoint of being decent egalitarian democrats. I believe it is impossible to establish rational grounds for giving the early-stage embryo less "moral status" than the later-stage fetus or newborn—without also dehumanizing, in principle, other classes of human beings, or making our humanity so conditional that the weak become more vulnerable to exploitation. If it is certain powers that the embryos lack, then there will always be other non-embryonic human beings that also lack them. And so the only rational view of the embryo that is fully consistent with democratic decency and democratic equality is the welcoming one—to treat the embryo as "one of us."

And yet, I am not fully convinced that it is "rational," or simply rational, to "love" an embryo as a neighbor. For it seems to require a profound, perhaps absurd, faith in reason itself to accept what our rational conclusions about the embryo might one day demand of us—especially if something like "therapeutic cloning" really worked, and if the choice before us were really the child who is dying or the embryo that might save him.

Perhaps the parent who allows his child to die—rather than create and destroy an embryo to save him—is an angel. Perhaps he embodies the highest, and hardest, love of all—the love that stings, the love of the cross. But he is also a monster, or will seem like a monster to most members of democratic society. He is, in a word, a "monstrous angel." This parent is perhaps the human type that most perfectly reconciles "unconditional love" and "conditional excellence." Perhaps he is the existential hero of conservative bioethics. But he is also the character least suited to our democratic society and democratic values, a society that produces neither angels nor monsters, and that was deliberately designed to produce neither angels nor monsters.

And so this is our paradox: a true commitment to democratic equality demands welcoming the early embryo as one of us. But the potential implication of this view, if embryo research or "therapeutic cloning" were to work as promised, is something no democratic society could accept, and for some good reasons.

A POLITICS OF THE EMBRYO

So what is a democratic society to do? It would be irresponsible to reflect on the larger meaning of the embryo debate for American civilization without saying something concrete about the actual policy dilemmas before us. To do so, we need to sort out the three overlapping issues of *abortion, embryo research*, and *new ways of making babies*—which are morally related but also morally distinct, and which are governed by very different legal regimes and different political realities. Abortion is the destruction of a developing human life, *inside the womb*, in the supposed interests of the carrying mother, and sometimes because the developing child has a genetic defect or is the "wrong" gender. Embryo research is the exploitation and destruction of embryos, *in the laboratory*, for the sake of medical advances and potential therapies. And our new techniques for making babies involve the creation, screening, and manipulation of embryos *in the laboratory*, with a view, in the future, to implanting these genetically tested, modified, or cloned embryos into the child-seeking mother. In the first case we have a developing life we do not want; in the second case we seek and then destroy nascent human life in the effort to save those we love; in the last case, we want a child that we could not otherwise have, or we want a child of a particular sort—cloned, screened, or enhanced.

Taken together, these three issues reveal the profound moral and legal contradictions that have taken shape over the last thirty years surrounding the beginnings of human life: We worry about manipulating embryos in a way that might lead to a new eugenics, while protecting the legal right to destroy embryos and fetuses for any reason at all. It is legally possible, at the state level, to ban all research on embryos outside the body—and even to treat such embryos, as Louisiana does, as "juridical persons"[7]—while other states provide significant funding to destroy them. Scientists say that embryos outside the body are not human because they cannot develop to term, while pro-choicers say that once we implant them in the very wombs where they might develop we cannot legally protect them. For years, we have been engaging in revolutionary new techniques of producing children in the laboratory, such as preimplantation genetic screening, with no regulation and often no prior experiments on animals, and recent studies suggest that there might be real dangers and real harms to the resulting children.[8] We have engaged in this great baby-making experiment with the apparent approval of most American liberals, who seem to care more about not treating embryos as subjects (and thus imperiling, as they see it, the right to abortion) than protecting the well-being of the children whose lives begin in the laboratory. And while the FDA limply says it can regulate cloning-to-produce-children, they can only do so by treating the cloned embryo as a "product" (like a drug) that might imperil the health and well-being of the mother.[9]

No doubt, in the years ahead, the embryo question will continue to divide us; it will remain a wedge issue in American politics. The question is whether we can, at the very least, set some reasonable limits on the exploitation of nascent human life, such as banning the creation of human embryos solely for research; and whether we can sustain the limits that already exist, such as the prohibition on harvesting human embryos in animal or artificial wombs in order to use them for spare parts.

For if we were to begin harvesting fetuses as means to our ends, it would become impossible to see ourselves as simply a mediocre democracy, transgressing a sacred mystery, rather than a nation whose limitless pursuit of health has made us morally mad. We already tolerate most late-stage abortion—that is true. But abortion is not a good we seek in the way deliberate research on fetuses one day could be. Very few people see abortion doctors as heroes, but many people already see the work of embryo researchers as heroic. In the end, it may be that we have already gone too far to turn back, and that our willingness to use the earliest embryos already makes us the kind of people that will cross every line in the name of progress, or science, or medicine, or health. But I do not yet believe it. We are corrupt, but we are not evil. We are decadent, but we are not morally dead.

If I could stop all embryo research before it really gets going I would do so, and if I could put the embryo back inside the body, I would probably do so. But I cannot, and at least for now, nobody can. We can, however, try to stop the worst horrors—and worst temptations—before they arrive, and without conceding in principle those moral truths that we can never fully embody in law. This is not heaven, but it is a society we can all still decently live in—where a few angels can still flourish, where the worst monsters are kept at bay, and where American civilization is still largely a source of moral pride.

CHAPTER FIVE

Our Genetic Condition

AS WITH EMBRYO RESEARCH, the latest advances in genetics confront us with the gravest matters of human life—including how we have children, how we understand our identity, and how we face sickness and death. And perhaps like no other area of modern science, the mapping and manipulation of the human genome inspires both dreams and nightmares about the human future. The utopians dream of a new age of perfect babies; the dystopians imagine a looming nightmare of genetic tyranny. But neither vision of the future is the best guide for living well with the dilemmas of the genetic present.

In thinking about the new genetics, we often commit two errors at once: worrying too much too early and worrying too little too late. When the gradual development of genetic technology does not seem as wonderful as the dream or as terrible as the nightmare, we get used to our new powers all too readily. Profound change quickly seems prosaic, because we measure it against the radically altered future we imagined instead of the complicated world we truly have. Our technological advances—including those that require overriding existing moral boundaries—quickly seem insufficient, because the human desire for perfect control and perfect happiness is insatiable. At the same time, scientists assure us that today's breakthrough will not lead to tomorrow's nightmare. They tell us that what we want (like new cures) is just over the horizon, but that what we fear (like human cloning) is technologically impossible.

The specific case of human cloning is indeed instructive for thinking about the new genetics more broadly. In the 1970s, as the first human embryos were being produced outside the human body, many critics treated in vitro fertilization and human cloning as equally pregnant developments, with genetic engineering lurking not far behind. James Watson testified before the United States Congress in 1971, declaring that we must pass laws about cloning now before it is too late.[1] In one sense, perhaps, the oracles were right: Even if human cloning did not come as fast as they expected, it is coming and probably coming soon. But because we worried so much more about human cloning even then, test-tube babies came to seem prosaic very quickly, in part because they were not clones and in part because the babies themselves were such a blessing. We barely paused to consider the strangeness of originating human life in the laboratory; of beholding, with human eyes, our own human origins; of suspending nascent human life in the freezer; of further separating procreation from sex. Of course, IVF has been a great gift for many infertile couples. It has answered the biblical Hannah's cry, and fulfilled time and again the longing of most couples to have a child of their own, flesh of their own flesh. But it has also created strange new prospects, including the novel possibility of giving birth to another couple's child—flesh *not* of my flesh, you might say—and the possibility of picking-and-choosing human embryos for life or death based on their genetic characteristics. It has also left us the tragic question of deciding what we owe the thousands of embryos now left over in freezers—a dilemma with no satisfying moral answer.

But this is only the first part of the cloning story. Fast-forward now to the 1990s. By then, IVF had become normal, while many leading scientists assured the world that mammals could never be cloned. Ian

Wilmut and his team in Scotland proved them all wrong with the birth of Dolly in 1996, and something similar seems to be happening now with primate and human cloning. In 2002, Gerald Schatten, a cloning researcher at the University of Pittsburgh, said "primate cloning, including human cloning, will not be in our lifetimes."[2] By 2003, he was saying that "given enough time and materials, we may discover how to make it work."[3] In 2007, researchers at Oregon Health Sciences University announced the successful cloning of primates, which has since been repeated by scientists elsewhere.[4] And today, leading laboratories around the world are eagerly—and confidently—at work trying to produce the first cloned human embryos for research. If they succeed, the age of human reproductive cloning is probably not far behind.

The case of human cloning should teach us a double lesson: beware the dangers of both over-prediction and under-prediction. Over-prediction risks blinding us to the significance of present realities, by focusing our attention on the utopia and dystopia that do not come as prophesied. Under-prediction risks blinding us to where today's technological breakthroughs may lead, both for better and for worse. Prediction requires the right kind of caution—caution about letting our imaginations run wild, and caution about letting science proceed without limits, because we falsely assume that it is always innocent and always will be. To think clearly, therefore, we must put aside the grand dreams and great nightmares of the genetic future to consider the moral meaning of the genetic present. And we need to explore what these new genetic possibilities might mean for how we live, what we value, and how we treat one another.

Humanly speaking, the new genetics seems to have five dimensions: (1) genetics as a route to self-understanding, a way of knowing ourselves; (2) genetics as a route to new medical therapies, a way of curing ourselves; (3) genetics as a potential tool for human re-engineering, a way of remaking ourselves; (4) genetics as a means of knowing something about our biological destiny, about our health and sickness in the future; and (5) genetics as a tool for screening the traits of the next generation, for choosing some lives and rejecting others. Let us explore each of these five dimensions in turn.

GENETIC SELF-UNDERSTANDING

The first reason for engaging in modern genetics is simply man's desire to know himself. Alone among the animals, human beings possess the capacity and the drive to look upon themselves as objects of inquiry. We study ourselves because we are not content simply being ourselves. We are not satisfied living immediately in nature like the other animals do. Food and sex alone do not satiate us. We do not accept the given world as it is; we also seek to uncover its meaning and structure. Modern biology, of course, is only one avenue of self-understanding, one way of asking questions. But it is an especially powerful and prominent way of seeking self-knowledge. Instead of asking who we are by exploring how humans live, the biologist asks who we are by examining the mechanics of human life. Genetics fits perfectly within this vision: it seems to offer us a code for life; it promises to shed empirical light on our place in nature; it claims to tell us something reliable about our *human* design, our *pre-human* origins, and our *post-human* fate.

But it is also true that the more we learn about genetics, the more we seem to confront the limits as well as the significance of genetic explanation. As the cell biologist Lenny Moss put it:

> Once upon a time it was believed that something called "genes" were integral units, that each specified a piece of phenotype, that the phenotype as a whole was the result of the sum of these units, and that evolutionary change was the result of new changes created by random mutation and differential survival. Once upon a time it was believed that the chromosomal location of genes was irrelevant, that DNA was the citadel of stability, that DNA which didn't code for proteins was biological "junk,"

and that coding DNA included, as it were, its own instructions
for use. Once upon a time it would have stood to reason that the
complexity of an organism would be proportional to the number of its unique genetic units.[5]

But in fact, the triumph of modern genetics has also meant the hum-
bling of modern genetics. On the biggest human questions, the permanent limits of genetics should have been obvious from the start:
Did we really think that the study of man's DNA would would explain,
as Leon Kass puts it, "what life *is*, what is *responsible* for it, or what it
is *for*," or that genetic knowledge would explain to anxious parents
why *this* child is born sick while *that* one is healthy?[6] But even on nar-
rower questions—such as the role of a person's genes in shaping his
behavior—the mysteries remain as great as ever. The study of genetics has opened up more questions about man than it has settled.

Yet even if we are relieved at discovering the limits of genetic
determinism and thus the possibility of human freedom, our hunger
for genetic explanation remains strong. Disease also imperils human
freedom, after all, and we can still hope that genetics might help us
conquer that mortal threat. We can still hope that genetics is the secret
of disease, if not the secret of life.

GENETIC THERAPY

And this points us toward the second dimension of the new genetics:
the search for medical cures. Modern science, unlike ancient science,
does not rest on the foundation of curiosity alone. It seeks to conquer
nature, not simply to understand nature's meaning. And while man
may be the only truly curious animal, his curiosity is not his only
guiding passion. He also seeks health and he certainly fears death. Like
other animals, human beings seek comfort and survival. But unlike
other animals, we possess the capacity to pursue comfort and survival
through the systematic application of reason. Modern science, especially modern biology, promises the "relief of man's estate," in Francis

Bacon's famous phrase, in return for the right to explore nature without limits. Descartes skillfully negotiated this bargain centuries ago, and I quote here a passage much cited by those interested in the origins of modern science:

> So soon as I had acquired some general notions concerning Physics ... they caused me to see that it is possible to attain knowledge which is very useful in life, and that, instead of that speculative philosophy which is taught in the Schools, we may find a practical philosophy by means of which, knowing the force and the action of fire, water, air, the stars, heaven, and all the other bodies that environ us, as distinctly as we know the different crafts of our artisans, we can in the same way employ them in all those uses to which they are adapted, and thus render ourselves the masters and possessors of nature.[7]

Not surprisingly, the "nature" we most seek to "master" is our own. We seek to conquer human disease, and perhaps even to make death itself a series of conquerable diseases. It is apparently part of our genetic code to revolt against our genetic fate.

Of course, the "speculative philosophy" of the Schools that Descartes sought to leave behind was religious metaphysics—which is to say, the search for man's place in the cosmological whole and before God. The new science and the old religion thus seem to present us with two different ways of revolting against our biological fate: The religious believer seeks such revolt *beyond nature* in God, by looking beyond our genetic deficiencies to the hope of eternal salvation. The scientist seeks such revolt *through nature* in science, by understanding nature's mishaps (or mutations) so that we might correct them. The unknowable God, if you believe He really exists, promises better long-term results; He "cures" us forever, but only after death. The empirical scientist, if you give him enough public funding, provides better short-term results; he cures us now, but only for a while. This does not mean that science and religion are enemies: religious people are often great scientists, and great scientists are often deeply religious. But it does suggest that the cure-seeking scientist lives on the narrow ridge between holiness and rebellion: He imitates the old God by healing the sick; or he supplants the old God by believing that he can eradicate all sickness, by working within nature rather than looking beyond it.

Genetics, in this sense, is simply a new frontier in the long ascent of modern medicine. It aims to repair broken genes or correct disease-causing mutations by direct intervention. And it aims to use our growing understanding of the human genome to diagnose and treat human disease with greater precision.

But it turns out that most diseases are more complicated than genetics alone, and that markers for identifying and predicting a given disease do not always or easily translate into usable knowledge about the disease's causation. The capacity to fix genes with perfect precision and without side effects is also proving remarkably difficult. Already, there have been some high-profile examples of gene-therapy trials going terribly wrong, and the field now proceeds with perhaps a more befitting caution. Over time, of course, there is little doubt that our genetic knowledge will improve modern medicine and thus prove a great blessing to us all. But there also seems little doubt that the new genetics will probably not be the therapeutic panacea that many once hoped, and which many scientists and policymakers offered as a central justification for the human genome project. Biological knowledge and biological control are simply not the same, even when it comes to curing disease, and most certainly when it comes to so-called genetic engineering.

GENETIC DESIGN

This brings us to the third dimension of the new genetics: the much-discussed prospect of designing our descendants—a prospect I find unlikely, certainly in our lifetimes. In the reproductive context, we already possess some (limited) capacity to pick and choose human embryos for implantation based on the genetic characteristics that nature gave them, as discussed more fully below. But this is significantly different from designing human beings with genotypes entirely of our own creation. By focusing so much energy on the dream and the nightmare of genetic engineering, we risk treating the real-life possibilities of genetic control as less profound than they really are. Yet again: we worry too much too early and too little too late.

To be sure, it may be possible to engineer various genetic monstrosities—like a human version of the monkeys with jellyfish genes that glow in the dark. Perhaps some modern-day Frankenstein will create fetuses with primordial wings; or children with seven fingers; or human beings that are part male and part female by design. If human life is seen as a mere canvas, and if the biologist sees himself as an artist thriving on "transgression," then genetic engineering is a real problem. And sadly, there is little doubt that someone, somewhere, will attempt such terrible experiments, and may succeed in producing at least embryonic or fetal monsters. But I also have little doubt that most democratic societies will pass laws that prohibit the biological equivalent of postmodern art. Precisely because it is so grotesque, such monster-making is not our most serious ethical problem.

Democratic societies, after all, do not seek the monstrous; we seek the useful. And the worst abuses of biotechnology may come in trying to make the seemingly monstrous dimensions of life disappear in the name of mercy, by screening and aborting those with handicaps or deformities that we believe make their lives not worth living. There will always be knaves who reject society's laws and principles and engage in monstrous acts for their own sake. But the real challenge is to consider those uses of genetic knowledge and genetic choice that are both technically feasible (as science, not art) and that seem to serve the humanitarian aspirations of liberal society. It is those potential abuses that have some utilitarian justification—such as improving life, or ending suffering, or guaranteeing every child a healthy genome, or expanding reproductive freedom—that we must confront most squarely.

But since many people worry so much about genetic engineering, one would be remiss to ignore it entirely. So let me offer a brief critique. The most tempting reason to engage in genetic engineering is to assert new kinds of control over our offspring, and to design children with certain desirable *human* attributes: children with high IQs, beautiful appearance, remarkable strength, amazing speed, and photographic memories. Some might even seek to design human offspring with better-than-human attributes. But these scenarios strike me as technically unlikely and humanly misguided. Technically, I doubt whether we will ever gain, or gain soon, the sophistication to engineer certain human traits *de novo*, and I doubt whether the traits we seek to engineer are so clearly rooted in a definable genetic pattern that we can deliberately replicate or improve the pattern. At the very least, I believe the project of trying to find such patterns and implement such designs would involve so many grotesque failures that the backlash would be swift.

More deeply, I doubt that we can actually design a *better* human being—even as a genetic thought experiment. If the goal of the designer is human *excellence* or better-than-human excellence, he must begin with an idea of excellence itself. And here, I think, we face two insurmountable hurdles: First, I doubt that we can improve upon nature when it comes to making a better musician, or artist, or scientist. It is hard to imagine a composer better than Mozart or a playwright better than Shakespeare. In seeking to maximize some human trait by genetic manipulation, we will most likely deform other crucial traits, and thus deform the excellent human wholes that nature so mysteriously and so remarkably supplies. And if we seek, say, to make faster men to run our races, have we really created better men—or just biological machines? Cars move faster than men; pitching machines throw harder than pitchers—but neither invention is better than human; they are merely sub-human things.[8] And even if we could make as many Mozarts as we like, do we really serve the cause of human excellence by making that excellence so common? I doubt it.

The final major barrier to the genetic engineering project is the fact that superior talent is not the only form of human excellence.

Many of the most admirable human beings do not live lives dominated by measurable achievement, but lives of fidelity, or charity, or love, or courage. Perhaps there are important genetic predispositions to such traits of character, but good genes are rarely enough to make good men, even if bad genes sometimes make individuals so psychologically impaired (or chemically imbalanced) that virtue is beyond their reach. Moreover, I suspect that even replicating these good genetic predispositions will be beyond the engineer's reach, because they involve so many biological factors that go beyond mere genetics. Even if our technology improved, I doubt that we can engineer more virtuous offspring—which is the only real measure of whether genetic engineering would make human life truly better.

All that said, the one form of "genetic engineering" that does demand our attention is the very real prospect of human reproductive cloning—a way of controlling the genetic make-up of our offspring with great precision, by copying the genetic make-up of someone already here. The ethical and social significance of human cloning is profound, involving a deep violation of the relationship between parent and child. But technically, cloning is remarkably simple compared to other imagined forms of genetic engineering. It does not involve manipulating the interlocking pieces of the human genome, but the wholesale replication of an existing genotype. It is more like copying a great novel already written than writing a great novel from scratch.

And it is this technical ease, in fact, that makes cloning a genuine worry, not simply a distracting dream or nightmare. Cloning involves a perverse form of self-love, by imposing one's own genome on one's son or daughter. It robs new life of an open-ended future, and it forces the young clone to live always and forever in the shadow of his elder genetic twin—in the shadow of both his past accomplishments and past failures. In the end, human cloning may prove a test case of our capacity to limit the dehumanizing uses of biotechnology, and our capacity to defend those human goods—like the family—that make human life truly human.

———————

GENETIC FOREKNOWLEDGE

But if most forms of genetic engineering, beyond cloning, are probably not in the offing, this hardly means that the new genetics is socially and ethically insignificant. Certainly not. What it means is that we need to pay much closer attention to the human meaning of *genetic knowledge* itself—both how we use it and what it does to us once we possess it. And this brings us to the fourth dimension of the new genetics: the meaning of gaining partial foreknowledge about our biological fate, and especially the meaning of knowing bad things (or good things) about our biological future.

Of course, to be self-aware at all is to have some foreknowledge of our mortal destiny: We know that death will one day take us; we know that natural disasters, or terrible accidents, or vicious attacks could make this day our last day; we know that some mysterious ailment could strike us without warning. Those of us who eat the wrong foods and spend too much time at our desks know that heart problems and clogged arteries may lie in our future; even without sophisticated genetic tests, we know about the presence of hereditary diseases in our families; and we all know that time will eventually win its final victory, whether at age 70, or 80, or 90, or 100.

And yet, most of us live our day-to-day lives without focusing too much on our own mortality. For better and for worse, we do not live each day as if it could be our last; we do not make the fact of death a dominant reality in our everyday lives. When a loved one dies or some tragedy strikes, we are perhaps reminded of our mortal condition; we might imagine our children throwing dirt into our graves. But the immediacy of life quickly returns, and we live again, for a while, as if the horizon of the future were very long, if not indefinite.

Strangely, modern individuals are both more obsessed with death and less aware of death than their less-modern forebears. We are obsessed with trying to avoid death through better diets and better medicine, yet we are less aware of death because it rarely strikes us in untimely ways, at least compared to the omnipresence of death in the lives of our ancestors. In modern societies, most people die after living full lives, not from mass plagues, or mass killings, or infant mortality.

In an essay on the meaning of mortality, Hans Jonas quotes the Psalmist asking of God: "So teach us to number our days, That we may get us a heart of wisdom."[9] His point is not primarily religious but existential. If we lived as if tomorrow were forever, we would lack the urgency to live boldly and love deeply. And if we believed that this life would last forever, even the sweetest things might become routine.

But in the age of genetic testing, the instruction to "number our days" takes on new meaning, since these tests may allow us—or force us—to number them with increasing precision. Today, there are numerous deadly diseases that we can diagnose through genetic testing with absolute or near-absolute certainty, and long before we experience any visible symptoms. For some of these diseases—like Huntington's—there is no cure; the diagnosis is a death sentence, giving the likely age of onset, the likely period of decline, and the likely age of death if nothing else kills first. For other diseases—like breast cancer—genetic tests can offer a highly reliable if not perfect indication of a person's susceptibility to the disease, with potential treatments ranging from preemptive surgery to remove one's breasts and ovaries to intense monitoring to detect the coming cancer as early as possible.

But does this genetic foreknowledge make life better or worse? Is there a case for genetic ignorance? At what age and under what circumstances should people know their genetic fate? These are hard moral questions with no easy answers.

In those situations like Huntington's where the diagnosis is clear and there is no cure, genetic self-knowledge seems like both a blessing and a curse. It is a blessing, because it might lead individuals to an uncommon wisdom about the preciousness of life; it might move them to live without wasting time, because they know just how short their time really is. And yet, such foreknowledge must also seem like a curse; the permanent presence of looming death might make living seem worthless; there are too many projects they know they can never finish and too many ambitions they know they can never fulfill. Their genetic death sentence may come to feel like a living death, with no escape except pharmacology or suicide.

In those situations where some therapeutic preemption is possible, like for those who test positive for the breast cancer mutation, the young often face drastic and wrenching decisions: Is the greater chance of longer life worth living with the scars of mastectomy, or living without the possibility of bearing children of one's own? Is it really better to have the knowledge that makes such a tragic choice necessary, rather than the ignorance that would allow us to live without being so haunted until the disease really comes?

Right now, the number of diseases we can test for genetically is somewhat limited, and many of these tests offer clear positive or negative diagnoses. But what may be coming is a world of imperfect knowledge about terrible possibilities—with a battery of tests that give greater and lesser probabilities of getting certain diseases, at certain times, compared to the general population. All of our human fears will be sharpened; our paranoia made more precise; our anxieties given a genetic scorecard. What good is this knowledge to us, especially when the power to diagnose will come long before the power to cure—the so-called "diagnostic-therapeutic gap"? And yet, will we be able to resist this new form of high-tech astrology? Will it teach us to number our days and make us wise? Or will it make life seem like a short trip through a genetic minefield—by forcing us to confront every morning the ways we might die?

GENETIC CHOICE

These types of genetic foreknowledge take on new meaning when we move to the reproductive sphere, and when the burden is not simply living with knowledge of one's own potential fate, but deciding whether such knowledge is a morally compelling reason to abort an affected fetus or discard an affected embryo. And this leads us to the final dimension of the new genetics: the use of genetic knowledge to make reproductive decisions, to decide between life worth living and life unworthy of life.

For a long time, we have worried about the so-called "enhancement problem," and feared that some people would use genetic technology to get an unfair advantage for their offspring. But this, I believe, is the wrong worry. The real danger is that the limitless pursuit of equal results—the desire to give everyone a mutation-free life, and thus an equal chance at the pursuit of happiness—will actually undermine our belief in the intrinsic equality of all persons. The pursuit of genetic equality will lead to the age of genetic discrimination. And in some ways, it already has.

Of course, if we could avoid conceiving a sick or disabled child, we would do so. And if we could safely cure Tay-Sachs or Down syndrome during pregnancy or *in vitro*, we should do so. But once conception has taken place, and in cases where there is no cure, we are left with the decision to accept or reject a life in-progress—a life that is real enough to us that we can evaluate and pass judgment on its genetic characteristics. With the arrival of preimplantation genetic diagnosis (PGD), we may face a radical transformation of assisted reproduction—a transformation made more significant by the rising numbers of women and families turning to IVF to have children. In this new world, genetic testing would become a standard part of IVF, and the tested embryos would be divided into different classes: those doomed to suffer killer diseases like Tay-Sachs would be separated from those that are not; those doomed to suffer disabilities like Down syndrome would be separated from those that are not; those prone to suffer late-onset diseases like breast cancer would be separated from those that are not.

By making reproduction into a process of division by class, we transform the welcoming attitude of unconditional love into a eugenic attitude of conditional acceptance. Of course, we would do this in the name of compassion, or mercy, or equality. We seek to give our children healthy genetic equipment, and to spare those who would suffer by "nipping them in the bud." But the pursuit of genetic equality requires a radical program of genetic discrimination. Whatever we might think about the moral status of the early embryos tested in P G D, they are certainly not nothing. They are real organisms, with the same genetic identity as embryos that they would have through life if those who created them in the first place decided to let them live.

Seen clearly, the real danger of the genetic age is not that the "gene-rich" will outpace the "gene-poor"; it is that the pursuit of genetic equality will erode our willingness to treat those who are genetically unequal as humanly equal. We will replace the hard work of human love for the disabled with a false compassion that simply weeds out the unfit. It is hard to see how the equal dignity of persons with Down syndrome is served by treating Down syndrome as a legitimate reason to abort. And it is hard to see how parents will experience pregnancy with any equanimity or joy if they have a full genetic read-out of their embryo or fetus, and must decide whether the mutation for breast cancer, or Parkinson's, or Alzheimer's disease is reason enough to abort and try again. This is the moral paradox at the heart of genetic control: In seeking an existence without misery or imperfection, we may make ourselves more miserable and imperfect, and we will be tempted to do miserable things in the name of a misguided mercy. In the very act of bringing new life into the world, we will already be thinking about how our future child will die.

BEYOND GENETICS

Without question, the advance of modern genetics is one of the great achievements of our time, an example of the creative and truth-seeking spirit at the heart of our humanity. But too often, we easily assume that the progress of science is identical to the progress of man. The truth, as always, is much more complicated. Many men and women of the past were superior in virtue to us now, and many scientific discoveries of the present and future will prove a mixed blessing, and sometimes even a curse.

The new genetics will deliver us many goods but also confront us with many burdens. We will need to make choices, and those choices will require philosophical judgments about "better" and "worse," not only scientific judgments about "possible" and "impossible." We will need to think especially about the goods in life that are higher than health—the goods that make being healthy worthwhile. And this is the very task that modern genetics is least equipped to handle.

We will also need to challenge the lazy assumption that genetic knowledge is simply "neutral," with a meaning that depends entirely on "how we use it." For this, too, is much too simple. New knowledge is never neutral; it is always a way of being in the world, a way of seeing our condition, a way of seeking truth, happiness, and virtue. Genetics is no exception, and genetic knowledge will never eradicate or eliminate those perplexities of life that require the kind of wisdom that no material science can ever offer.

CHAPTER SIX

The Commerce of the Body

THE SPIRIT OF modern capitalism is as varied as the souls of modern men. Virtually every human type finds his place in the modern economy, and virtually every human desire is eventually transformed into a commodity. The cosmetic surgeon specializing in breast implants. The observant Jew rushing to finish work before sundown. Sex stories on MTV and salvation stories at the movies. Oil-drilling corporations and embryo-destroying start-ups. *Queer Eye for the Straight Guy* and NASCAR racing. Precisely because modern capitalism is rooted in the idea of human freedom, men and women are free to produce and consume in ways that reflect their own private ideas of the good life, both for better and for worse.

In general, the modern economy involves decent men and women working hard to better themselves and provide for their families. But commerce sometimes goes deep into the human gutter—the multibillion-dollar pornography industry is perhaps the grossest example —dragging many ordinary people down with it. The modern economy relies largely on average people doing average work, competently if not brilliantly. But it also stirs the ambitions of those who seek to remake the world with their talents and visions, often with technologies that aim to satisfy the full spectrum of human hungers.

Perhaps the most striking dimension of the modern economy is the new commerce of the body, including an impressive array of new biotechnologies and biological procedures that promise to improve, control, or manipulate our native biology. In myriad ways, the better body is for sale—from anti-impotence drugs to anti-depressants, from cosmetic surgery to low-carb diets, from baby-making clinics promising you a healthy child to the current push to legalize the buying and selling of human organs. And if one looks ahead to the biotechnologies of the future—improved mood- and memory-altering drugs, stem-cell-based medicine, genetic muscle enhancements, new techniques for controlling the genomes of one's offspring—it is clear that the commerce of the body will only become more ambitious, selling bodily perfection to anyone with enough disposable income.

This leaves us to wonder: Is "bio-capitalism" something radically new, bringing with it a new spirit and new dilemmas? Or is it simply the continuation of modern capitalism's promise to better our condition indefinitely? No doubt the answer is some combination of continuity and novelty. The interesting question is whether the novel dimensions of bio-capitalism are so fundamental that we need to rethink our moral intuitions about capitalism itself. In short: Does the new commerce of the body portend a moral crisis for modern capitalism?

From the beginning, the idea of modern capitalism was connected to various notions of the good life, or different assessments of the best life possible for limited, selfish, and imperfect human beings. Morality and modern commerce were always inseparable, and the defense of commerce (like the lament) was originally made in moral terms. In the modern age, three different attitudes toward life and commerce are especially worth considering: the spirit of *God-seeking enterprise* embodied in early Protestantism; the *irreverent self-love* embodied in the likes of Voltaire; and the *worldly moderation* best articulated by Adam Smith. To be sure, typologies such as this one often distort as much as they clarify; history is messy and complex, and the history of capitalism is winding and tumultuous, with passionate defenders, savage critics, and many unexpected turns. Still, the presence of these three different spirits of capitalism is undeniable, and undeniably important. To understand where we are heading, we need to revisit where we came from.

CAPITALISM'S THREE SPIRITS

In *The Protestant Ethic and the Spirit of Capitalism*, Max Weber describes how a new idea of salvation—a new creed about the relationship between man and God, worldly life and other-worldly grace—unexpectedly initiated the age of modern capitalism. It would be a vast oversimplification to say that there was a single Protestant Reformation; for there were many cross-currents, as Weber describes. But two ideas in particular—Luther's idea of "calling" and Calvin's idea of "predestination"—fundamentally altered the behavior of believing Christians and the trajectory of the West. Worldly work could now be understood in vocational terms. "The fulfillment of worldly duties is under all circumstances the only way to live acceptably to God," describes Weber. "It and it alone is the will of God, and hence every legitimate calling has exactly the same worth in the sight of God."[1]

But only by combining the idea of "calling" with the theology of "predestination"—that is, the belief in salvation by God's unfathomable grace alone, breathed into us at birth—did the spirit of capitalism find its paradoxical roots. For men could not live in practice or for long with a grace so mysterious, or with the state of their eternal souls so uncertain. They wanted "proof"—proof to themselves, proof before others, and proof before God that "I" am indeed saved. This desire for proof gave believing Protestants an "irrational" will to work with little interest in savoring the worldly fruits of their labors. The individual toiled instead as a sign of other-worldly salvation, and in accordance (as he saw it) with a divinely chosen calling. Every detail of life was rationalized and perfected; even the smallest sign of waywardness might be a sign of one's un-chosenness. Practical science was welcomed and mystical speculation discouraged.

The result of such an ethic, according to Weber, was a magnificent increase in material wealth, due largely to the accumulation of capital that came from producing so much and enjoying so little, from the combination of restless toil and ascetic self-denial. But the wealth produced as the outward fruit of man's piety threatened to undermine the inner commitment to God. As John Wesley, the founder of Methodism, declared:

> I fear, wherever Riches have increased (exceeding few are the exceptions) the essence of Religion, the mind that was in Christ, has decreased in the same proportion. . . . For the Methodists in every place grow diligent and frugal; consequently they increase in goods. Hence they proportionably increase in pride, in anger, in the desire of the flesh, the desire of the eyes, and the pride of life. So, although the form of Religion remains, the spirit is swiftly vanishing away. Is there no way to prevent this? This continual declension of pure Religion? We ought not to forbid people to be diligent and frugal; we *must* exhort all Christians, to gain all they can, and to save all they can; that is, in effect, to grow rich! What way then . . . can we take that our money may not sink us to the nethermost hell? There is one way, and there is no other under heaven. If those who *gain all they can,* and *save all they can,* will likewise *give all they can,* then the more they gain, the more they will grow in grace, and the more treasure they will lay up in heaven.[2]

Whether Weber is quite correct about the historical connection between the Protestant ethic and the birth of capitalism is a complicated and much disputed question. What seems clear is that God-seeking Protestants were central to the first flourishing of modern commerce, and that over time the fruits of such enterprise came to seem more desirable in themselves. As Weber put it: "The intensity of the search for the Kingdom of God commenced gradually to pass over into sober economic virtue; the religious roots died out slowly, giving way to utilitarian worldliness."[3] In this way, the formula was reversed —not material success as proof of salvation, but salvation via our material success.

Through Protestantism, commerce was made a realm of "grace." But over time, it was not God's grace alone, or at all, that men sought, but the grace of being a "self-made man"—the privileged condition that was secured by one's own labors or ingenuity, not bestowed as a divine gift and obligation. Before Protestantism, salvation was largely set apart from (or above) the realm of commerce—in the sacraments, the monastery, or the Sabbath. Protestantism weakened this separation—directing men, if somewhat unintentionally, to see the labors of life as proof of salvation, and eventually as its very source. But sooner or later, the self-made man confronts the limits of his own self-made grace. He is struck by misfortune, or boredom, or mortality. His grace is haunted and incomplete. The "religious roots" of commerce continue to lurk as ghosts within the modern economy. To this day, we are still haunted by the salvation that modern commerce once promised, and still hunting after the kinds of salvation it might yet give us—not in heaven, but in the flesh.

For Voltaire, the delights of the flesh were worth celebrating, and he admired commerce precisely for its capacity to promote worldly goods (including bodily pleasures) through freedom and exchange. Where the Protestant ethic prized self-denial, Voltaire celebrated self-love; and where the Protestant believer labored out of devotion to a saving God, Voltaire celebrated commerce for making such pious devotions irrelevant. "Religionists may rail in vain," he wrote. "I own, I like this age profane." He liked its physical comforts and the room it afforded for his playful, "worldly mind." He led a life of wild speculation—filled with financial schemes that would have made the managers of Enron proud—and he praised the London Exchange as a place where the only "infidels" were those who went "bankrupt."[4]

As Jerry Muller describes in his superb book *The Mind and the Market*, the real enemy for Voltaire was religious enthusiasm, which led men to slit each other's throats over archaic and trivial superstitions. "Here Voltaire is the prophet of the profit motive," Muller describes. "Compared to the competitive quest for salvation, the quest for wealth is more likely to make men 'peaceful' and 'content.' Compared to the altruistic crusade of forcibly saving one's neighbor's soul, even if it leaves his body in ruins, the pursuit of wealth is a potentially more peaceable pursuit, and one that leaves one's neighbor content."[5]

But it was not just religious conflict that Voltaire abhorred, but the pious man's devotion to a false salvation, his idealization of a wretched past at the cost of making a better future. In a poem with the fitting title "The Worldling," Voltaire pays tribute to the wonders of his age—the "needful superfluous things," the "luxury and pleasures." He mocks Adam and Eve for the wretchedness of their flesh, the dirty ground they slept in, the tasteless food they ate.

> My fruit-eating first father, say,
> In Eden how rolled time away?
> Did you work for the human race,
> And clasp dame Eve with close embrace!
> Own that your nails you could not pare,
> And that you wore disordered hair,
> That you were swarthy in complexion,
> And that your amorous affection
> Had very little better in't
> Than downright animal instinct.
> Both weary of the marriage yoke
> You supped each night beneath an oak
> On millet, water, and on mast,
> And having finished your repast,
> On the ground you were forced to lie,
> Exposed to the inclement sky:

Such in the state of simple nature
Is man, a helpless, wretched creature.[6]

Eve, in other words, could use a trip to the perfume counter and the salon. To embrace such a "wretched creature" is to be nothing more than an animal. Beyond the flesh, Voltaire praises the artists and the architects—the real makers of "grace." He delights in what is visible to the eye inside "rich golden frames," not what is knowable to the soul oriented toward heaven. The poem ends with Voltaire's fitting words of self-praise: "Terrestrial paradise is where I am." He is a "worldling" and nothing else, living in a paradise of "self-love" and "happy commerce," one that he desperately hopes to sustain—decaying flesh be damned.

Adam Smith, you might say, offered a moderate vision—between the Protestant quest for other-worldly salvation and Voltaire's irreverent delight in the luxuries of the flesh. With Voltaire, he believed that an alternative needed to be found to the wars of religious piety, and that state-regulated salvation was a recipe for tyranny and slaughter. And yet, he did not see religion itself as an enemy, and he took for granted, as Irving Kristol and others have argued, the habit-forming effects of traditional institutions like church and family.[7] Without the Puritan work ethic, it is unlikely that Smith's practical vision would have gotten off the ground. But Smith did not offer a commerce of salvation—worldly or other-worldly—but a commerce of progress, one that expanded man's liberty and gradually improved his condition. It was a sober and practical vision for sober and practical men. He was interested in building a decent society, by taking seriously both man's rational self-interest and his capacity for self-restraint, both his natural acquisitiveness and his latent civility. And he sought a society that improved the condition of all willing individuals, not simply a society where the strong triumphed over the weak, or where the wealthy pursued life's niceties while the poor remained in a condition of permanent desperation.

Smith, in other words, sought to build a future that "worked"—and by most accounts, he succeeded tremendously. We live in the world he built, with souls still shaped in large measure by his vision, and a politics still informed by his realism about the limits of radically remaking the human condition by conscious design. Smith's "system of natural liberty"[8] worked in two basic ways: First, it explained how the natural desire for self-improvement and the range of natural human capacities could cohere to produce a prosperous economic system—one in which individuals responded to the changing needs of the market and the changing possibilities of production, and lived with the freedom to "better their condition" according to their own lights. In this way, the desire for private profit could serve the public interest, and the largely free market could produce an organic order from below, one impossible to create from above. Second, Smith showed how commercial life could have a civilizing effect on acquisitive individuals, who needed to work hard and tolerate others in order to prosper. The commercial society does not stamp out selfishness or spread the gospel of brotherly love. But it does channel self-interest and promote civil society among individuals with different backgrounds and tastes. And it creates the wealth necessary for somewhat higher aspirations, if not necessarily the desire to pursue them.

Smith believed an economic system should be judged in moral terms—judged for the kind of people it produces and the way of life it allows to flourish. And he was not blind to capitalism's moral shortcomings—including the rise of scheming businessmen moved only by greed and devoid of conscience, and the existence of laborers made dull and brutish by performing a few simple functions without end. But the problems of greed, nastiness, and stupidity were hardly unique to modern capitalist life, and in many ways they were much worse in pre-capitalist societies. The problem, of course, was and remains the limits of human nature itself; a social system, at best, could promote virtues and curb vices—not make average men into philosophers or saints.

Peculiarly, the one thing that pious Protestantism and Voltairean atheism agreed upon was that commerce was a good thing. Traditional (Catholic and Orthodox) Christianity had thought otherwise, and modern radical progressivism thinks otherwise, too. But in the formative moments of the modern age, the pious and the irreverent agreed about the virtues of enterprise. Adam Smith, in his wisdom, sought to assuage the struggle between them by focusing on this point of agreement; he believed that commerce was the way to peace in modern times. The age that followed was almost bound to be defined by commerce, because those who fought to shape it agreed almost only in their veneration of trade. But trade, alas, is not the most venerable thing, because what men buy and sell cannot address man's deepest longings, even if the culture of the marketplace sometimes curbs humanity's worst excesses.

BIOTECHNOLOGY AND THE COUNTERCULTURE

In 1991, with the last vestiges of communism crumbling and the Cold War ending, Irving Kristol warned that the greatest threats to a capitalist future were spiritual and cultural. "In a sense," he said, "it is all Adam Smith's fault. That amiable, decent genius simply could not imagine a world where traditional moral certainties could be effectively challenged and repudiated. Bourgeois society is his legacy, for good and ill. For good, in that it has produced, through the market economy, a world prosperous beyond all previous imaginings—including socialist imaginings. For ill, in that this world, with every passing decade, has become ever more spiritually impoverished."[9]

For all his great successes, Smith's ultimate error was his lack of "eschatological realism."[10] Man is not simply an average being who seeks to improve in material ways. He is also an imperfect being who yearns for perfection, a mortal being who yearns for immortality, and a suffering being who yearns for redemption, whether through divine grace or through his own technological conquest of nature. And so Adam Smith's world of practical commerce—a great success—is still haunted by the Protestant desire for other-worldly grace and by Voltaire's desire for "terrestrial paradise." We demand that material progress offer salvation—which is exactly what socialism once promised and what biotechnology may promise in the future. Or we demand that material progress be abandoned in the name of salvation—soberly, by those who seek to preserve sacred retreats in a profane world, or radically, by extremists who seek to dismantle modern life altogether.

This quest for salvation can either go "with the grain" or "against the grain" of modern commercial society. Modern science—especially modern biological science—has long gone with the grain: seeking useful inventions, practical advances, and the "relief of man's estate" through a growing mastery of nature's laws and human biology. Technology has long been the art of self-improvement, and commercial society has long been inseparable from the creation and dissemination of new technologies. The implementation of Francis Bacon's vision has rarely shocked the Smithian mind, and mostly pleased it.

By contrast, modernist culture—art, literature, mores, and manners —has largely gone against the grain of modern commercial society. It saw the bourgeois world as boring, repressed, and unsatisfying—a world of "one-dimensional men,"[11] hungry for property, ruled by old-fashioned values left over from outdated religions. Modernism sought a life of the spirit in a life of *immodesty*—a life without limits, sexual or otherwise. It saw the mass of men as automatons, and it saw mass society as guilty of the degradation of both nature and culture. At the same time, it imagined that the autonomous individual was a creature without shame, a being beyond sin. It believed that alienation was a problem of history, not a condition of our nature. This attitude was epitomized in the counterculture of the 1960s—with its liberation of the body from old taboos, and its childish illusions about the remaking of man.

Back then, it seemed as though the culture of technology and the counterculture were mortal enemies. The machine vs. the spirit. Dionysian feeling vs. rational investigation. Gradual progress vs. spontaneous liberation. And of course, in some ways, they were and still remain bitter enemies. But perhaps not in the most important ways.

For it may be that the peculiarities of our own recent history mask a deeper connection between the counterculture and the culture of modern technology—a connection grounded in the belief that human limits should be overcome, taboos are anathema, and human shame is an illusion. Both cultures believe that no knowledge or no experience should be off-limits, and that death is an unfair or unnecessary sentence—to be overcome by technology or mocked artistically into submission. Both are willing to go where modest men never went before, at least not in public.

Let me make this point with a rather unpleasant example. Within a few months of one another, the following two stories appeared: Story #1, in *The New Yorker*, was about new works of art dealing with the human form. It included a picture of a grotesque sculpture, consisting of a number of naked children, connected to one another in the flesh, with penises as noses.[12] Story #2 was about a promising new technique of assisted reproduction, which allows women to remove a piece of their ovaries, freeze it indefinitely, and implant it into their arm or abdomen as a source of eggs whenever they decide to have children.[13]

Now I suspect many bourgeois scientists would find the penis-faced statue appalling, though they might defend the right to produce it as freedom of expression, akin to their own freedom of research. But the artists, I suspect, would actually admire the scientist's biological "transgression," the splicing of reproductive organs out of their "normal" context, the making public of once private parts. And even if the scientists reject such works of art as absurdities, modern biotechnology—and much else about modern commerce—has benefited greatly from the triumph of postmodern culture. For it was the radicals of the 1960s who cleared away the very taboos surrounding the body that would have inhibited the newest biotechnological possibilities. Can we imagine the commerce of the body today—or even the science that underlies it—without the prior triumph of the culture of immodesty? Would there have been terrain upon which scientists—and their investors—feared to tread if the counterculture had not tread there first, knocking down all moral barriers? Could it be that scientific rationalism and post-modern irrationality have more in common than it once seemed?

The genius of commerce is that it tames remarkable things; it makes past transgressions seem normal. What shocks the parents bores the children—both in culture and in science. Living together before marriage, test-tube babies—that's yesterday's news. We can already imagine a future where cosmetic surgery is as common as orthodontics; where mood-altering drugs are a mass phenomenon, like vitamins (or painkillers) for the soul; where people sell their deceased loved one's organs; and where 10 to 15 percent of women reproduce using in vitro fertilization, screening their embryos for sex, height, and other desirable genetic predispositions. And we are left to wonder: What will it be like to live in such a world, to raise a family in such a world, to work in such a world, to invest money in such a world? What will be the relationship between biotechnology, morality, and commerce?

THE NEW COMMERCE OF THE BODY

Of course, most biotechnology is admirable; it is a continuation of bourgeois progress as we have long known it, whose only negative effect is raising our medical expectations much more and much sooner than it can actually meet them, healing some patients even as it stirs resentment in those it cannot yet cure. But there are also reasons to believe that the new commerce of the body is growing increasingly removed from Smith's sober recognition of human limits. It promises perfection, not progress; and it heeds no limits, treating the sacred and the profane as indistinguishable objects for sale, ruled only by the amoral law of supply and demand. Lest this all seem too abstract, consider a few everyday examples.

EXAMPLE 1: THE BETRAYAL OF THE CHILD

By now, the idea of selling one's eggs or sperm to others who wish to produce a child is commonplace. One need only look in any elite college's newspaper to find advertisements offering substantial sums of money—$25,000, $50,000, or even $100,000—for egg donors with perfect figures and high SAT scores. There are numerous companies that specialize in brokering eggs, often catering to very particular tastes. A law-student friend of mine, for example, received the following solicitation in the mail:

> Dear Potential Egg Donor: The Genetics and IVF Institute is looking for healthy, college-educated, ethnically diverse women between the ages of 21 and 32 to assist infertile couples by becoming an anonymous egg donor. . . . You will be adequately compensated for each cycle you complete . . . beginning at $5,000 [and going] up to $45,000. . . . Help an infertile couple experience the joy and fulfillment of parenthood.

Now in market terms, this potential transaction makes perfect sense —matching a willing seller and a willing buyer. Each party gets what it desires—tuition money for the seller, the seeds of a new child for the buyer—and no one is coerced into anything. But what is the human meaning of what is happening? A couple desperately seeks a child of its own, a child biologically related to the father genetically and the mother by pregnancy. This is why IVF came into existence in the first place—because the infertile seek not just a child to raise, but a child who is flesh of their flesh. But to make this possible, in some cases, they need a seller who is willing to abandon her own biological child; willing to be an anonymous donor; willing never to set eyes upon the child that is flesh of her flesh. The buyers who desperately want a biological child need a seller who sees having a biological child as no big deal. In market terms, again, this makes sense: a case of two parties valuing different commodities differently. But in human terms, it means finding a seller who denies the very human longing that the buyer wishes to act upon. It requires a seller who is willing to betray his or her own flesh and blood offspring—not out of desperation, but for a price.

EXAMPLE 2: THE SHAMING OF THE FATHER

By now, ads for anti-impotence drugs are common fare in magazines and on television. One of the most memorable campaigns starred the former Chicago Bears coach "Iron" Mike Ditka—the consummate tough guy, who takes the "Levitra Challenge" to "stay in the game." Coach Ditka is apparently comfortable discussing his erectile dysfunction, and perhaps proud of his continued desire for virility. He flaunts his nakedness—the loss of his powers, the hunger for his powers—for all the world to see, including his children.[14]

Now consider another story—the story of Noah in the book of Genesis, naked in his tent; and the story of his sons (Shem and Japheth), who so revere their father that they do not look upon him. They walk backward to him and cover him with their cloak. As Leon Kass describes: "They intuitively understand that, were they to see with their own eyes their father's nakedness, their family order would be permanently altered. . . . By protecting Noah's dignity and authority, they safeguard their own capacity to exercise paternal authority in the future. . . . They knowingly choose to live leaving some things in the dark, without pressing back to the naked truth about temporal beginnings or ultimate origins." Even in his old age, they see their father as a giant, the source of their own being.[15]

Today, by contrast, we leave nothing in the dark and we strip down every giant. Both proper pride and proper shame are thrown to the wind. While Coach Ditka might seek such drugs in the name of his manliness, it is precisely his manliness that is compromised. By flaunting his desire to "stay in the game," he loses the reverence—the majesty—that a dignified old man should command of those beneath him. Instead, he lays out his nakedness for public consumption, including the sons who now cannot help but shame him. In his quest for potency, he reveals his ultimate dependence, with no cloak to preserve any ennobling illusions.

EXAMPLE 3: THE MODERN BIRTH-MARK

A few years ago, the Fox network aired a "reality" show called *The Swan*, which took a score of average-looking women, sent them to a team of cosmetic surgeons who remade their bodies under the knife, and then put the refurbished ladies on stage to decide who is the most beautiful—to decide which ugly duckling is now the swan. Already, cosmetic surgery is no longer simply the province of actors in Hollywood and politicians in Washington. It is becoming—slowly but steadily—a mass phenomenon, and perhaps soon a middle-class phenomenon. Some parents now give their teenage daughters nose-jobs and breast implants as high school graduation presents.

As a consequence, physical beauty is no longer seen as nature's endowment but as man's creation. Aging is no longer accepted gracefully but fought back with the knife. Imperfection is increasingly intolerable. Like Georgiana in Nathaniel Hawthorne's "The Birthmark," we subject ourselves to technicians of the body in the hope of being loved, or in the hope of making bystanders into lustful worshippers of our flesh. And the question is: In so doing, what have we lost? After all, beauty is never an achievement but an undeserved gift of nature. Why does it matter whether the giver of the gift is God, the gods, or the master surgeon? What is lost in removing this year's imperfections? Perhaps nothing. The trouble, of course, is next year's imperfections. And the problem is that our new "look" will inevitably change all our pre-surgery relations: To our parents, perhaps it will be a partial indictment of their own sub-par appearance; to our spouse, perhaps it will be an admission that I was not beautiful enough then; to our children, perhaps it will teach them that they too might need cosmetic surgery someday, since the genes they inherit come from the pre-surgical self. Even as we remake the flesh in accordance with our will, we cannot escape the attachments of the flesh that we did not will—the attachments to our parents and our children. We have the swan's face and the ugly duckling's family. In seeking a more perfect figure or visage, we shame the very people we think we love.

EXAMPLE 4: THE BROKEN SOUL

Of course, it is not just the body we seek to fix but also the embodied mind. Commercials for mood-altering drugs are ubiquitous, and the use of such drugs has skyrocketed in the last decade, with distracted two-year-olds to the depressed elderly to everyone in between seen as part of the market.

The commercials for these drugs almost always work in the same way: a troubled child or employee—failing at work, failing at school, growing more distant from loved ones. Then a drug that promises, as one slogan puts it, to reveal "the real you." And then a sudden transformation, a new life of smiles, friends, and productivity. A thirty-second *commedia* with neurochemistry as the playwright. Without question, such drugs can help many individuals who suffer from terrible mental illness, rooted in chemical problems in the brain, that only medication can ameliorate. For such people—the truly sick—psychotropic drugs are a godsend. And no doubt the strategy of selling these drugs is the same as selling any other product: convincing people they are inadequate as they are, yet within reach of perfection; making people feel sick and desperate, only to discover that what they lack is some liberating product.

But surely something deeper is at work here, when the inadequacy is the psyche itself, and the liberation involves, in part, a new identity altogether. The real questions about the rise of psychotropic drugs are too significant to deal with here—questions about why so many people feel so depressed in the first place, why they believe only medication can help them, and who they really are once they start taking these mood-altering medications and start forming human relationships that depend on taking the drugs to sustain them. I can only note here the strangeness of this new marketing of dependence, and the significance of coming to believe that life's dilemmas are fundamentally problems of brain chemistry, only solvable by medication. Perhaps we will also come to believe the inverse: that life's best possibilities are likewise matters of chemistry, only achievable with medication.

In a certain sense, of course, this is all true: we live as given bodies, with drives that we do not fully control and cannot fully explain, and limits that come with our particular set of DNA. But we also live—or have long lived—with the belief that we are more than our chemicals, that our choices, joys, and miseries are more than inexplicable neuroactivity, that there is a difference between what is real and what is induced.

Perhaps the deepest problem with such drugs—taken by a widow to ease the pain of her mourning, or after a terrorist attack to calm one's sense of horror—is that they will confound, not restore, our sense of the world as it really is. To sleep easily amid carnage or rest easily after the death of a beloved spouse is to live in a world of fantasy. It is to seek salvation by no longer being fully human, with all the ups and downs that our condition entails.

EXAMPLE 5: THE EMBRYO AND THE COFFEE GRIND

My final example is somewhat more futuristic, but only somewhat. Depending on where the science takes us, it is not too far-fetched to imagine that human embryos will one day be valuable medical commodities—harvested routinely as a source of stem cells. Embryo destruction for research purposes is now commonplace. Scientists are already exploring methods that would allow us to produce human eggs artificially, thus eliminating the only practical barrier to embryo production on an industrial scale.[16] And no doubt such embryos will trade in the market like any other commodity—perhaps even on the "commodities exchange."

Perhaps I exaggerate, but it is an exaggeration with a point. What the market does is veil the meaning of what it uses so that everything can be used efficiently. It tames the remarkable and makes it seem normal—like everything else. It reduces each commodity to measurable data—where what matters is not the different things in themselves but the differential movements on the chart: coffee grinds up, embryos down; computer parts up, body parts down; Viagra up, Paxil down. Even the individual who is troubled by this prospect—who still asks whether a human embryo deserves more respect than a natural resource—will find it hard not to participate: Will he reject embryo therapies that might save his child? Will he leave his job at the health insurance company that covers such therapies? Will he sell the mutual fund that buys shares in an embryo-production company?

We should not forget that the goal of embryo commerce would be humanitarian—the pursuit of health, the very good that modern societies most desire. But the means are, arguably, a form of cannibalism of the weak by the strong—if a cannibalism not obvious to the eye because embryos look so un-human, and thus without a visceral repugnance to awaken our conscience and guide our behavior. But the violation is no less real for being unobvious, and it is only possible because we now take for granted a truly remarkable thing—the power to initiate human life outside the body, the power to see and hold what was once left shrouded.

And this, I think, is what we should most fear about biotechnology's transformation of modern capitalism: that in the desire for worldly salvation—salvation of the flesh—we will profane the sacred, with the modern marketplace greasing the skids. We will come to believe that bio-capitalism can sell us everything we desire, and thus come to accept that everything is for sale.

THE MORAL LIMITS OF CAPITALISM

Such a critique is not meant to show ingratitude for our economic prosperity and freedom. Only a fool would belittle the genuine virtues of progress, and I can imagine no better way to organize a modern society than democratic capitalism. At the same time, however, we must face up to the fact that modern commerce is often a moral problem, the capitalism of the body most especially.

Perhaps ironically, it is the conservative friends of commerce who will most likely see the profaning power of commerce. Critics on the left mostly attack capitalism because they want more of the very things that capitalism creates, but believe "big business" is keeping the fruits of progress from the "little guy." But conservatives realize that the deeper problem with capitalism is that it creates many things we should not create in the first place, and may ask us to do many things we should not do at all.

Without turning our backs on the modern economy—a prospect as foolish as it is impossible—we need to reconsider the relationship between modern technology and modern commerce, in the hope that we can salvage Smith's moderation from Bacon's excesses, and perhaps salvage better answers to man's permanent questions than simply buying what the cosmetic surgeon and the neurochemist eagerly want to sell us.

CHAPTER SEVEN

A Jewish-Catholic Bioethics?

THE TERM "JUDEO-CHRISTIAN" has entered our civic vocabulary for good reason. On many of the deepest issues of human life—the meaning of sex, the dignity of the family, the creation of human beings—Jews and Christians stand together against the secular image of man. But occasionally, even close friends have disagreements. In a March 2005 essay in the online magazine *Slate*, William Saletan observed that traditional Catholics and conservative Jews do not always think alike when they gather at meetings like those of the President's Council on Bioethics.[1] According to Saletan, Catholics raise deep questions and then presume to answer them with divinely confident reason. Jews raise those same deep questions but seem less certain that reason can ever finally settle them. Catholics oppose clear evils like embryo destruction. Jews worry about diffuse evils like the corruption of our sensibilities.

There's some truth in Saletan's claim, though matters are, of course, much more complicated. The particular Jews he discusses—Leon Kass, Charles Krauthammer, Yuval Levin, and even me—are hardly representative of Jewish bioethics. In many respects, we are outcasts. We oppose most or all forms of embryo research, for instance, and vehemently oppose the creation of embryos solely for experimentation and destruction. By contrast, with all the division among the different branches of Judaism—about keeping Kosher, intermarriage, driving on the Sabbath—destroying embryos for research is a point of remarkable theological agreement. The preeminent Reform, Conservative, and Orthodox Jewish organizations in America have all given their ethical endorsement, seeing embryonic stem-cell research as not only permissible under Jewish law but an embodiment of Jewish values. Reverence for life means seeking cures for disease; *ex vivo* embryos are a justified sacrifice—or little sacrifice at all—in the sacred cause of medicine.

A few prominent Jewish ethicists and halakhic experts dissent, seeing embryo destruction as potentially a prohibited form of feticide. But these voices are in the Jewish minority. Most Jewish thinkers support embryo research with few qualms, and many Jews see opposition to embryo research—or even the denial of federal funding for such research—as an illegitimate imposition of Christian values.

JEWISH SOURCES AND JEWISH WISDOM

Rabbi Elliot Dorff is a typical example. His guidelines on embryonic stem-cell research—adopted nearly unanimously by Conservative Judaism's Committee on Jewish Law and Standards—begin by describing, at great length, the cutting edge of stem-cell science: the various methods and sources for deriving embryonic stem cells, the potential to test new drugs and develop new cellular therapies, and the state of research at different American laboratories. The document revels in its scientific sophistication before turning to the fundamental ethical question: Should Jews support the destruction of human embryos for research?[2]

To answer this, Dorff turns to Jewish law on abortion, and especially the Jewish understanding of what embryos and fetuses are as they develop. After forty days, he says, the fetus is classified by the ancient rabbis as "the thigh of its mother"; before forty days, he says, the embryo is "simply water."[3] Dorff says that it makes sense to follow such teachings only if they cohere with the truths of modern science. And then, inexplicably, he concludes that they do, ignoring the significance of what we now know biologically: that a new organism exists from the moment of conception; that the very first cell divisions are orderly and purposeful; that forty days is a meaningless moment from the standpoint of continuous embryological development; that by forty days the primordial head, arms, and legs have already formed, the primitive heart tube is present, the nerves of the face are developing. The notion that "simply water" is the best metaphor for understanding the unfolding human being in our care is absurd. It is morally and theologically irresponsible to seek the fruits of modern science in the form of stem-cell research without confronting the facts of modern embryology in order to understand what embryos really are.

Jewish thinkers such as Rabbi Dorff commit two errors simultaneously. They embrace modern biomedical science as a faith in itself, and thus lose the mystical vision that might allow them to see embryos as more than simply microscopic cells—to see embryos as God might see them. And they appeal—often selectively—to ancient religious sources without confronting the new scientific facts that make some of these sources a problem. They are, at once, too attached to modern biology and too removed from modern biology. And one wonders whether some of Dorff's confusions—like describing gametes and embryos without distinction as "potential building blocks of life," or equating embryos with other cells of the body, or describing the destruction of an embryo as "taking a part of an object"[4]—are not deliberate efforts to make embryo research seem more innocent than it is.

Perhaps the problem is simply that most Jewish thinkers have chosen one Jewish value—the good of healing—as the prism through which to see the old sources. Other considerations—like the law against deliberate killing or the belief in the sanctity of every life as created in the image of God—might lead to different conclusions. The Jewish sources themselves pull in many directions: the *Zohar* declares that "he who causes the fetus to be destroyed in the womb . . . destroys the artifice of the Holy One. . . . For these abominations the Spirit of Holiness weeps."[5] R. Meir Simchah says that the killing of a fetus is punishable by "death at the hands of heaven."[6] According to the opinion of *Ba'al Halaahot Gedolot*, the Sabbath may be violated to save an unborn child, even in the first forty days of development.[7] *Sanhedrin* 57b interprets the biblical text "Whoso sheddeth the blood of man, by man shall his blood be shed" to include "Whoso sheddeth the blood of man, within man shall his blood be shed." And who is a "man within man," the rabbis ask? "A fetus within the womb."[8]

To be sure, there are many Jewish sources that cut in other directions, especially in the standing they accord to embryos in the first forty days. But these sources deal mostly with the laws of purity for potentially pregnant women or for women who miscarry; they do not deal directly with the moral meaning of deliberately killing early embryos. Likewise, the key text in Exodus, which requires a man who causes a miscarriage by colliding with a pregnant woman to pay a monetary fine, does not deal with the meaning of deliberate killing.[9] Deliberate killing, however, is what embryo research necessarily requires, especially research that creates embryos solely for exploitation and destruction.

Some Jewish thinkers, including Dorff, argue that the embryo *ex vivo* has limited moral standing because it cannot develop to term outside the womb. But surely all human beings deprived of the environment they need to flourish have "limited potential for life." A bird trapped in a cage may never learn to fly, but it is no less a bird for the harm we caused by putting it there.[10] A grown woman without food or water will surely die, but this lack of sustenance does not make the doomed person less than human. If anything, it challenges the humanity of those who left her there to die in the first place.

Perhaps the one great halakhic exception to the pro-embryo research consensus in modern Judaism is Rabbi J. David Bleich, a giant of ethical and legal scholarship in the Orthodox Jewish world. Bleich rejects the argument that embryos and fetuses *in utero* possess greater moral standing than embryos and fetuses *ex utero*. And he reminds Jews to stand more humbly before the mystery of new life by reminding them of Ecclesiastes: "As thou knowest not what is the way of the wind, nor how the bones do grow in the womb of her that is with child; Even so thou knowest not the work of God Who doeth all things."[11] And Bleich concludes his meditation on the ethics of stem-cell research by praising the Catholic Church for its witness in defense of nascent human life: "The Catholic Church now uniquely fulfills a different role in the transcendental divine plan, i.e., it tenaciously promulgates the notion of the sanctity of fetal life and the teaching that abortion constitutes homicide. Non-Jews who engage in that endeavor do so with divine approbation. Non-Jews engaged in fulfilling a sacred mission are surely deserving of commendation, applause, and support."[12]

Bleich is not convinced, as Catholics are, that early embryos are the moral equivalent of full human persons. Neither are some of the Jewish conservatives involved with the President's Council on Bioethics. Bleich's questions are grounded in the mystery of Jewish sources, while the former chairman of the council, Leon Kass, sees a possible tension between our moral intuitions about early embryos as worthy of great respect but different from fully-formed human beings, and the rational account of early embryos as full persons from conception to death. He opposes embryo destruction but is not convinced that *ex vivo* embryos are necessarily "equal" in moral standing to post-embryonic human beings.[13] My own view is that the Catholic arguments are indeed the most rational, but accepting them in a moment of trial—such as choosing between the child who is dying and the embryo who might save him—would require a faith that is truly other-worldly and thus seemingly absurd to this-worldly eyes.

But whatever fine philosophical differences may exist in theory, Jewish conservatives who engage publicly on these issues have spent the past several years fighting for prohibitions on embryo destruction. As Jews, we know well what it means to treat some human lives as less than human, or some human beings as there for experimentation. We know the moral hazards of justifying such dehumanizing violations on the grounds that embryos are "going to die anyway," just the way some Nazi doctors justified their inhuman experiments. Embryo destruction is not the moral equivalent of the Holocaust; to make such a direct comparison is grossly irresponsible, for it fails to capture the horror of seeing one's children or parents sent before your eyes to the gas chamber, and the full-scale barbarism that is required to systematically slaughter men with human faces like one's own. But the moral lessons of the Holocaust should give us the wisdom to oppose making embryo destruction the new foundation of modern medicine. That, it seems to me, is what Jewish ethics demands.

VISIONS OF HOLINESS

But it is also only part of the story. In the post-Holocaust age, the Jewish mind is not only keenly aware of the dangers of mistreating innocent life. Jews are also afraid of the demographic death of the Jewish people. The pathos of infertility—a continual theme in the Hebrew Bible—is more powerful than ever, while the place of procreation remains central to the Jewish idea of holiness, to Jewish self-understanding as a sacred people, to the Jewish obligation of passing down God's way from one generation to the next. As it says in the Talmud:

> "Should the number of Israelites happen to be two thousand and two myriads less one, and any particular person has not engaged in the propagation of the race, does he not thereby cause the Divine Presence to depart from Israel?" Abba Hanan said in the name of Rabbi Eliezar: "He deserves the penalty of death; for it is said, 'And they had no children, but if they had children they would not have died.'" Others say: "He causes the Divine Presence to depart from Israel; for it is said: 'To be a God unto thee and to thy seed after thee'; where there exists 'seed after thee', the Divine Presence dwells among them; but where no 'seed after thee' exists, among whom should it dwell? Among the trees or among the stones?" [14]

With this in mind, recall again how human embryos came to exist outside the body at all, to be seen with human eyes and held with human hands. We produced embryos outside the protective darkness of the womb in order to give those whom nature made barren a natural child. In vitro fertilization was a technological answer to Sarah's laughter and Hannah's cry: "O Lord of Hosts, if You will look upon the suffering of Your maidservant and will remember me and not forget Your maidservant, and if You will grant Your maidservant a male child, I will dedicate him to the Lord for all the days of his life."[15] To give birth is to be eternally remembered; to be childless is to be eternally forgotten.

Many orthodox Jews see this as grounds for defending even reproductive cloning in certain situations if it were possible and safe. As Yitzchok Breitowitz, one of Orthodox Judaism's finest ethicists and halakhic experts, has argued: "Let us assume that the individual is the last survivor of a family that was decimated in the Holocaust and let us assume that he was castrated in a concentration camp. If he dies, there will be no perpetuation of his family line. . . . Cloning is as close as the couple in this scenario could possibly come to producing a child that is on some level the genetic product of both of them. One clearly could defend, I think, the morality of the use of reproductive cloning for that limited purpose."[16] To some degree, every Jew after the Holocaust feels like a Holocaust survivor. And while Breitowitz explores the many dilemmas raised by reproductive cloning, he believes that cloning to produce children presents no inherent problem from the standpoint of Jewish ethics and Jewish law. The wisdom of engaging in human cloning should be judged case by case, he says, and surely not banned by the state.

And so here we have yet another great Jewish irony and internal Jewish conflict: For decades Leon Kass has been the most eloquent and passionate opponent of human reproductive cloning, seeing it as a violation of the dignity of human procreation, a violation of the relationship between the generations, a violation of the uniqueness of new human life, a violation of the sexual character of human reproduction.[17] And yet, it is Orthodox Jews who make the ethical case for cloning with greatest force—as a way to perpetuate a holy people when it is the only biological alternative.

The Jewish defense of cloning strikes me as woefully misguided—a deep misunderstanding of what it means to participate as husband and wife in the creation of new life. But the Jewish defense of cloning is a perversion of something more genuine: the special meaning of procreation within Judaism, and what it means not only for the human family in general but for *this* particular human family. As Kass himself has explained, children are part of the Jewish answer to mortality, a Jewish way of participating in the immortal. Children connect Jews directly to Abraham, Isaac, and Jacob—to those blood relatives chosen by God to bring God's new way into the world. In prayer, Jews sing *l'dor v'dor*: "from generation to generation." In law, Judaism passes down through birth, not baptism. This is why in vitro fertilization finds nearly universal support within all branches of Judaism—as permissible, never obligatory—and why even those Jews who oppose embryo research often reluctantly defend in vitro fertilization.

Catholics, in their more universalistic wisdom, do not. They oppose in vitro fertilization precisely because it corrupts rather than fulfills the dignity of human procreation; because it separates the unitive and procreative purposes of marriage; because it turns the mysterious birth of new life into a technological project; because it paves the way for the age of human cloning and genetic engineering; and because it destroys thousands of embryos as "byproducts" and abandons thousands more as "spares." Most deeply, the practice of in vitro fertilization compromises the connection between sex and holiness: the way the sexual encounter of man and wife, created in the image of God, gives them a glimpse of the divine communion of the Triune God and the mysteries of Creation itself.[18]

The good that in vitro fertilization has produced—the many lives now living—is undeniable. But so are the moral hazards. As a universal ethic, the Catholic position is compelling, certainly in its prudence about the many evils in vitro fertilization has already caused (like mass embryo destruction) or will likely facilitate in the future (like growing genetic control over our offspring).

But as a particularistic ethics—as a practice engaged in by God's Chosen People, confronted with the suffering of Sarah and Hannah—in vitro fertilization *may* have a theological purpose and thus a moral justification, if done within marriage and without producing excess embryos. It may be that Jews and Catholics—who share so much in their understanding of the dignity of procreation and marriage—must part ways in their understanding of what holiness in action requires in certain tragic cases of infertility.

As a Jew, I respect the Catholic position deeply and tremble at the practice of initiating new life in the laboratory, even as I wonder at the magnificence of giving new life—flesh of the couple's flesh—when in vitro fertilization is the only way to do so. And I hope that Catholics tremble when they tell an infertile couple—including an infertile Jewish couple—that having a biological child should not be done even if it could be done, even as they wonder at the magnificence of the Catholic vision of human sexuality and its connection to the mystery of God's inner life. This is a disagreement among friends mutually devoted to holiness; it is a disagreement that God, if He exists, will surely settle in His own way and in His own time. But on most things that count—including embryo research—faithful Jews should stand alongside their Catholic friends as Judeo-Christians, opposing together the imageless image of man that secularism offers.

PART THREE

From Generation to Generation

CHAPTER EIGHT

Why Have Children?

OVER THE PAST few years, a new "demographic crisis" has emerged as a subject of intense debate: the most affluent, most advanced, freest societies of the world are not having enough children to sustain themselves. Recent books—including Phillip Longman's *The Empty Cradle* and Ben J. Wattenberg's *Fewer*—have described the potentially tragic consequences of this fertility decline.[1] Lamenting the collapse of modern birthrates, world leaders as diverse as Vladimir Putin and Pope Benedict XVI have advocated pro-natalist state policies.[2] Popular magazines and newspapers that once worried about the horrors of a "population explosion"—mass starvation in developing countries, environmental catastrophe, the subjugation of women trapped by the excessive burdens of serial motherhood—today ask whether free societies mean to perpetuate themselves at all.

Right now, the answer, with a few exceptions, is no. The numbers are indeed staggering. Since the 1950s, the total fertility rate (TFR) in Europe has fallen from 2.7 to 1.38[3]—an astounding 34 percent below the replacement rate of 2.1, which is the average number of children per couple needed for a society to sustain itself. Japan's fertility rate is 1.23, and its average age is already 43.5 years and climbing.[4] (The world average, by comparison, is in the mid-20s.)[5] A large number of nations, including Russia, Spain, Italy, South Korea, and the Czech Republic, have TFRs between 1.0 and 1.3[6]; some of these nations (most notably Russia) are already experiencing rapid population decline.[7] Generations of children are growing up without brothers or sisters, and a sizeable percentage of men and women in the most advanced nations will never have any children at all.

Compared to most of its democratic peers, the United States is still in decent demographic shape, with a fertility rate hovering near replacement, albeit with sizeable differences from region to region (higher fertility in most "red states," lower fertility in most "blue states") and between child-rearing immigrants and child-avoiding natives. But like every other advanced nation, the U.S. is also heading toward a mass geriatric society, with more elderly dependents and fewer grown children to care for them or grandchildren to replace them.

The consequences of the birth dearth now worry people of every imaginable political, religious, and ideological stripe, though for a range of different reasons. One major set of worries, widely shared, is economic. In a 2004 study commissioned by the European Union, the RAND Corporation warned that as the percentage of the population in the workforce shrinks, countries experience reduced productivity, overburdened pension and social insurance systems, and a decreased ability to care for a needy and growing elderly population.[8] In other words: fewer workers, more retirees, and a fiscal crisis for the European welfare state. The economic problems do not stop there. Older populations are less likely to be innovative and entrepreneurial, and less likely to produce the consumer power necessary to drive national economies. Moreover, those states that raise taxes on the young to support programs for the old will only make it more difficult for the rising generation to afford children of their own. The result is a vicious cycle of economic stagnation, a graying of society on the way to decline or extinction.

But the deeper demographic worries are cultural. To Longman, the central looming problem is what he calls "the return of patriarchy." Since religious fundamentalists are still having children while liberal secularists are not, Longman fears a "new Dark Ages," a demographic reversal of the Enlightenment in which zealous Christians at home and radical Muslims abroad will eventually inherit the earth. He therefore wants liberals to become pro-natalist, and urges democratic societies to enact child-friendly social and economic policies. If children are more affordable, he hopes, happiness-seeking adults will have more of them.[9]

George Weigel, relying heavily on Longman's data, flips his argument on its head. In *The Cube and the Cathedral*, he argues that the deepest cause and most tragic consequence of population decline is the abolition of Christian Europe, the birthplace of human rights and human progress. He believes that Christian renewal is the best hope for saving the West from the twin dangers of secular nihilism's soul-destroying barrenness and radical Islam's nation-destroying fecundity.[10] A similar argument has been advanced by the columnist Mark Steyn, who attributes the modern West's low fertility to its "lack of civilizational confidence." The "design flaw of the secular social-democratic state," Steyn writes, "is that it requires a religious-society birthrate to sustain it."[11,12]

Other cultural dangers loom as well. Once the welfare states of Europe, Japan, and the United States are in crisis, euthanasia may seem like the most rational, or perhaps the only plausible, solution. Once the childless generations grow old, they will face their mortality without children to care for them, comfort them, and mourn them. The childless freedom of the past will often end in isolation. The old will die with no one there to say "Kaddish" (the Jewish prayer of remembrance), and with little assurance that the faith of their fathers will persist *l'dor v'dor*, from generation to generation, once they return to the pleasureless dust from which they came. Perhaps the fact that having a child to say Kaddish seems to matter so little, or only matters too late, lies at the heart of our cultural and demographic problems.

Of course, one must always tread lightly in judging the choices of free people in free societies—choices often made for understandable human reasons. To appreciate our predicament in all its complexity, we need to probe more deeply than the demographers and economists typically do; we need to explore the human motivations and longings that lie beneath the statistics. Why are free people not having children, and will the fertility bust usher in the decline of the West that many people now fear? What is new about our contemporary patterns of birth and death, and what lessons can we learn from the cultures of the past? Only by asking these questions can we attempt to answer the most important question of all—Why have children?—in a way that might resonate with modern people who do not necessarily believe that the God who created the heavens and the earth also commanded them to "be fruitful and multiply."[13]

PATRIMONY, PLEASURE, AND PERPETUATION

Fertility control is hardly a new phenomenon in human history, and the current age is not the first in which the fear of depopulation aroused grave economic and cultural concern. "In our own time," wrote Polybius in roughly 150 B.C., "the whole of Greece has been subject to a low birthrate and a general decrease of the population, owing to which cities have become deserted and the land has ceased to yield fruit." The reason for this demographic crisis, he believed, was the cultural decadence of the age. "For as men had fallen into such a state of pretentiousness, avarice, and indolence that they did not wish to marry, or if they married to rear the children born to them, or at most as a rule one or two of them, so as to leave these in affluence and bring them up to waste their substance, the evil rapidly and insensibly grew."[14]

In both ancient Greece and Rome, limiting fertility was seen as the prudent course for established families—a way to preserve wealth from generation to generation, undivided by multiple heirs, and a way to preserve each man's share of immortality by preserving the family name. Ancient forms of birth control were almost surely a fertility-lowering factor, and the desire to control the number and quality of one's offspring was defended by poets, philosophers, and aristocrats alike. As Hesiod described, one would "hope for an only son to nourish his father's house, for this is how wealth waxes in the hall."[15] Aristotle suggested that "the proper thing to do is to limit the size of each family, and if children are then conceived in excess of the limit so fixed, to have miscarriage induced before sense and life have begun in the embryo."[16]

In general, Greek and Roman children were not seen as sacred gifts but as products of nature—sometimes wanted and sometimes not, sometimes with the potential for human flourishing and sometimes sub-human, sometimes useful to their progenitors and sometimes liabilities. "Monstrous offspring we suppress," wrote Seneca, "and we drown infants that are weakly or abnormal."[17] The natural affections of mothers and fathers for their children, surely not absent, were usually governed (at least among elites) by the patrilineal ideal or the pursuit of pleasure, sexual and otherwise. As Angus McLaren explains in his superb study, *A History of Contraception*, "the Roman elite did not relish the prospect of their urbanized, civilized style of life being jeopardized by a horde of infants."[18]

But the good of the family and the happiness of the individual did not always serve the public good. At some point, the Roman combination of patrimony and pleasure broke down, prompting fears of depopulation and giving rise to laws that punished the unmarried and the childless and publicly rewarded those with three or more children. In support of such "marriage legislation," the emperor Augustus supposedly read from the famous speech of the censor Quintus Caecilius Metellus Macedonicus, delivered in 131 B.C.:

> If we could survive without a wife, citizens of Rome, all of us would do without that nuisance; but since nature has so decreed that we cannot manage comfortably with them, nor live in any way without them, we must plan for our lasting preservation rather than for our temporary pleasure.[19]

But natality for the state did not win universal favor or assent. "I shall beget no sons to swell Rome's glory; not of my son shall historians tell," wrote one Roman poet to his mistress. "Let me be your one joy; you at my side, I have no need of sons to feed my pride."[20]

Making the leap from Athens and Rome to ancient Jerusalem, one encounters another male speech about childlessness—the speech of Elkanah to Hannah, his beloved but barren wife. "Hannah, why weepest thou? And why eatest thou not? And why is thy heart grieved? Am I not better to thee than ten sons?" Of course, we know that Elkanah already has sons and daughters of his own by another wife, which may account for his rather sanguine attitude. And perhaps his failure to understand Hannah's misery—"forgotten" by God and cruelly mocked by her fecund rival—reflects a distinctly male blindness to her distinctly female longing for a child. But whatever the cause, Hannah herself, unlike the Roman poet, could never say "I have no need of sons," and her ultimate reason for wanting them was the very opposite of "feeding her pride." Indeed, she sees the child she wants as a child wanted for God. She relies upon God to open her womb, and God Himself relies upon the fruit of her womb to sustain His holy way in the world against the "wicked [who] perish in darkness."[21]

The biblical idea of procreation is captured succinctly in the famous passage from Genesis: "Be fruitful and multiply." When God addresses Noah after the flood, His speech begins and ends with these words— first as a divine *blessing*, then as a divine *commandment*. God seeks to replenish the world destroyed by the flood, and to introduce a new chapter in the history of the world, ruled by men who are themselves ruled by God's moral law. Yet man, unlike the other animals, must be *commanded* to have offspring. Even as his God-like possibilities depend upon his animal-like capacity to reproduce, man alone among the animals has the power to reject and control his procreative drive. Man alone is tempted by illusions of self-sufficiency, or is prone to allow either present goods or present despair to curb his desire to raise up the next generation. One might even say that the Bible here offers a preemptive critique of sociobiology—that is, the idea that the passing on of our genes is the controlling impulse of human existence—before this new science was fashionable (and before it was clear that sociobiology cannot explain the current age of fertility decline). The sex drive is primordial, an ineradicable part of human nature; the procreative drive requires a commandment, bidding men to remember the future rather than lose themselves in the ecstasies or miseries of the present.

The idea of procreation as an act of devotion took a rather different form in ancient Sparta, where children were seen, from the beginning, as servants of the state's glory—and especially as future warriors or future mothers of warriors. "As a woman was burying her son," recalls Plutarch, "a worthless old crone came up to her and said: 'You poor woman, what a misfortune!' 'No, by the two gods, a piece of good fortune,' she replied, 'because I bore him so that he might die for Sparta, and that is what has happened, as I wished.'" The son, in the Spartan mother's eyes, was only her child if he a lived a life worthy of the state's glory, and of course some children failed. "Away to the darkness, cowardly offspring, where out of hatred/ Eurotas does not flow even for timorous deer./ Useless pup, worthless portion, away to Hell./ Away! This son unworthy of Sparta was not mine at all." [22]

Our ancient forbears thus set before us three different ideas of procreation, each of them corresponding to permanent human desires that often come into tragic tension. The first is *patrilineal pride*—the desire to preserve one's name immortally in the flesh of one's children, but to limit their number for the economic benefit of parent and child alike. The second, going beyond the notion of limiting the number of one's offspring, is *freedom from children*, to pursue instead the pleasure that needy infants make impossible, especially when their incessant cries are seen as disruptions of sexual bliss rather than its fruits, or as impediments to ambition rather than the reason for working. The third is *perpetuation as an act of devotion*—whether to the city, to God, or to the young themselves, who will inhabit a future one hopes to sustain but will never see. In its most elevated form, procreation as fidelity sustains the good life from one generation to the next; in its most perverted form, it treats the young simply as servants of the state's glory, forever ready to be sacrificed. These three desires—for patrimony, for pleasure, for perpetuation—still rival one another for our deepest devotion, today as yesterday. They are the permanent possibilities, the eternal reasons to have or avoid children, that persist in the human soul despite the many economic and technological revolutions that drive history forward.

WHY NOT HAVE CHILDREN?

When it comes to human procreation, the permanent possibilities are today shaped by certain historically unprecedented conditions. First, there is no stigma attached to being childless; a woman's worth, in this life or the next, is not demeaned by the dominant culture if she chooses a life without children. And unlike ages past, when the hardest physical labor required male strength and the most elite professions were restricted to men, the professional ambitions of modern women meet virtually no external restraints. Second, children are no longer economic assets for their parents, as they generally were in rural and early industrial societies; rather, they are economic burdens, voracious consumers who produce virtually nothing until their late teens or early twenties. The middle-class child, as Longman argues, is a million-dollar expense. Third, fertility control is now both uncomplicated and virtually absolute. Those who want to avoid having children can easily do so—without restraining their natural sex drive, without putting themselves at physical risk, and without resorting to infanticide or abortion.

Children are thus culturally optional, economically burdensome, and technologically avoidable. But having the *option* to avoid children is not a *reason* to avoid them; and for those with reasons to have children, the economic burdens seem bearable enough. So the question remains: Why do so many men and women in the most affluent societies in history seem to want so few offspring?

A small literature has been devoted to this question by now. In a 2005 essay called "What Do Women Really Want?," social scientist Neil Gilbert develops an attitudinal typology running from so-called "traditionalists"—that is, women with three or more children who "derive most of their sense of personal identity and achievement from the traditional childrearing responsibilities and from practicing the domestic arts"—to, at the other end of the spectrum, "postmodern" women who are childless "by choice" and focused on themselves and their careers. In the middle are "modern" women with one child and "neo-traditional" women with two children—ways of life that vary in degree, not in kind, from the big-family traditionalists and no-family postmodernists. Over the past few decades, Gilbert finds, the trend toward the "modern" and "postmodern" end of the spectrum has been significant, with predictable demographic results. In the United States, the number of women over the age of 40 who have three or more children dropped from 59 percent in 1976 to 29 percent today. During the same period, the number of women with no children has nearly doubled to 18 percent, and the number with one child (now 17 percent) is climbing faster than the number with two children (now 35 percent). In Europe and Japan, the figures are skewed even more heavily toward childless and one-child families.[23]

To be sure, for some men and women, childlessness is an un-chosen misery. But most childlessness in our age is clearly self-chosen sterility. Many childless individuals, especially women, eventually regret never having children. But most see their choice, at least in the long decades when having children is still an option, as the positive pursuit of an individualist, feminist, or environmentalist ideal, or even as an act of charity toward the unborn, since the current age is so miserable that giving life is seen as an act of cruelty.

The early stirrings of today's "childfree movement" have been traced by Elaine Tyler May, in her book *Barren in the Promised Land*, all the way back to the time when the great postwar baby boom in the United States was coming to a gradual end.[24] She cites a playful 1957 article titled "Pity the Childless Couple":

There's nothing sadder than the childless couple. It breaks you up to see them stretched out relaxing around the swimming pools ... all suntanned and miserable ... or going off to Europe like lonesome fools. It's an empty life. There's nothing but more money to spend, more time to enjoy, and a whole lot less to worry about.[25]

By the late 1960s and 1970s, these "musings of a good mother—on a bad day" would become an entire philosophy of life—a radical turn against motherhood, a defense of sex and romance unencumbered by children, an embrace of feminine ambition and worldly power. "For many women," wrote Ellen Peck in her 1971 book *The Baby Trap*, "the birth of children marks the end of adventure, of growth, of sexuality, of life itself."[26] The modern feminist eschews Hannah's longing, instead pursuing the Roman poet's freedom; she has "no need for sons to feed her pride." In her view, motherhood extinguishes femininity; giving birth to new life is a kind of premature death.

This redefinition of the meaning of life has now survived three decades in which feminism itself has been redefined over and over again. Today, for example, "choice feminism"—the idea that every woman should decide for herself the best mix of motherhood and career—is widely acknowledged to have failed, amid much bitterness on the part of women who complain they were deceived into thinking they could "have it all." In a widely-discussed essay in *The American Prospect*, the feminist Linda Hirshman contends that women who have tried to balance work and family end up sacrificing the former to the latter, living lives tyrannized by diapers and dependent on men for their sustenance. Instead of such false "choices," she advocates a return to feminism's radical roots—ruthlessly ambitious, focused on self and money, uninterested in children. If a woman must have a baby, Hirshman writes, she should stop at one. (Somewhat flippantly, she concedes that "if you follow this rule, your society will not reproduce itself.")[27]

In their case against children, feminists of this stripe find a passionate ally in the environmentalist movement. Like feminism, environmentalism is a cause with many faces, but a majority of its adherents subscribe to the basic maxim that the fewer people there are, the healthier the planet will be. In this self-negating view, man is not the measure of all things, and self-control—including reproductive self-denial—is sometimes necessary for the good of the natural world. The classic statement of this view is Paul Ehrlich's 1968 tract *The Population Bomb*, which declared: "We must have population control at home, hopefully through a system of incentives and penalties, but by compulsion if voluntary methods fail. . . . Population control is the only answer."[28] Ehrlich's most dire predictions—such as mass starvation around the world—obviously did not come true, but he has become only more certain in recent decades that drastic fertility reduction alone can avert a looming ecological catastrophe, for which America and the West will deserve the greatest part of the blame.[29]

In *Maybe One: A Personal and Environmental Argument for Single-Child Families*, Bill McKibben articulates this environmentalist sensibility with far greater moderation. He describes the human longing for children and the joys of being a father himself, and yet defends the moral obligation of modern families to limit their number of offspring in order to prevent an ecological meltdown. He describes his own painful decision to have a vasectomy after the birth of his only daughter, and argues that single-child families are not only environmentally virtuous but often best for the child, who gains a monopoly on her parent's love.[30]

As McKibben's formulation suggests, however, environmentalist zeal in itself is rarely the reason that people decide not to have children or to limit their progeny to one. It is, instead, a way to add a patina of self-sacrifice to decisions made on other grounds—mostly self-regarding or family-regarding ones. Not that such grounds are necessarily trivial or unworthy in themselves. Raising children is a labor, requiring a constant outlay of time and energy, imposing a constant burden of new and growing expenses, and limiting one's personal freedom for two decades or more. To many, the economic or psychic costs may well seem too high, and there is no shortage of experts supplying data to confirm that impression. As one *New Yorker* writer put it, "Married couples with children are twice as likely as childless couples to file for bankruptcy. They're seventy-five percent more likely to be late paying their bills. And they're also far more likely to face foreclosures on their homes. Most of these people are not, by usual standards, poor. They're middle-class couples who are in deep financial trouble in large part because they have kids."[31] According to another study, parenthood is often associated with higher rates of depression.[32]

No wonder, then, that many people decide to have just one child, or maybe two. For many parents—not themselves excessively selfish, or ideologically committed to childlessness, or ruled by ambition alone —the most compelling reason not to have more children is to benefit the child they already have. They are the modern democratic versions of the old Roman patriarchs—seeking not to preserve the family wealth they already possess, but to ensure that the next generation gets the best medical care, attends the best schools, and lives in the nicest neighborhoods. At its worst, this parental desire to give each child the best things possible can deform into a desire for (only) the "best child"—a misplaced quest for perfection through procreation and child-rearing, treating one's only son or daughter as the means to fulfill one's own private hungers. At its best, however, having only one child, or maybe two, is an effort by parents to act responsibly in a world of high economic expectations. Anti-children in effect, they are pro-child in intention.

THE CRISIS OF THE WELFARE STATE

Yet therein also lies the social dilemma: The one-child family may flourish economically, but, as we have seen, a society of one-child families can lead to "market failure" on a disastrous scale, especially when, as in our own system, the young are expected to pay for the old through various entitlement programs like Social Security and Medicare. Private wealth can lead to public impoverishment.

Admittedly, the relationship between demographics and economic prosperity is a complicated matter. In dysfunctional nations that still lack market economies, keeping fertility low is a form of short-term humanitarianism, reducing the extent of general misery. Moreover, it is quite possible that dramatic fertility reductions in countries like India and China have had beneficial economic effects, by limiting the number of dependent children, freeing women to join the "productive" sector, and temporarily expanding the working-age population. But even in such societies, this "demographic dividend" will eventually need to be repaid as the population ages and the ratio of young workers to elderly dependents reverses itself. When that happens, these aging societies will either need to produce more children (short-term dependents but future workers) or impose the one-child logic of brutality at the end of life instead of at the beginning, by euthanizing the old when they can no longer support themselves or contribute to the state.

In America, Europe, and the world's other advanced democracies, this same demographic shift—more elderly dependents, fewer workers, the overall graying of society—is already here and getting worse. Today's remedial measures—such as increased daycare benefits in Sweden and tax incentives for couples with children in France—are modest to a fault: too small to convince potential parents to have children (or more children), and likely only to increase the costs of already ballooning entitlements.

And today's worries are mild compared with what is almost certainly coming when the baby boomers retire. True, those who have only one child or no children are usually wealthier throughout the course of their lives than those who bear the economic costs of raising the young, and so they should have ample capacity to save for their own retirements. True, too, childless societies in which people accumulate assets during youth, spend them down when they are old, and leave nothing behind can perhaps manage a smooth transition to extinction. Alternatively, one can imagine a renewal through immigration, with young workers born in the heyday of African or South

American fertility moving to North America and Europe to make a living for themselves and to support the graying natives.

But merely to state such "solutions" is to reveal that they are partial at best. After all, the inability or unwillingness to see oneself as old and in need of care, or to envision a world that will continue after one is gone, might inhibit rather than enhance the accumulation of adequate resources to pay for old age, just as it inhibits the creation of "human capital" in the form of children. While some middle-class couples with two incomes and no children (or one child) would be able to pay for their own health care until death, using money their neighbors have spent to send their three children to college, many will not think adequately about the needs of old age until it arrives. And they will almost certainly exert their oversized political power to preserve their personal entitlements when it comes time to collect them. As for immigration, despite America's track record of success in integrating newcomers, deep reservoirs of discontent have already gathered on this matter, as recent debates have demonstrated. Meanwhile, the democratic nations of Europe, with little success in integrating immigrants and with expanding and radicalizing Muslim populations, feel a need to close their doors at the very moment when their economic survival requires them to remain open.

Ironically, even a sudden upswing in European and American birthrates would not offer an immediate answer to today's demographic crisis. Children enter the world as helpless infants, not as high-tech entrepreneurs or geriatric nurses. At the same time, state programs intended to encourage higher fertility or to ease the economic burden of raising children are very expensive and will only worsen the fiscal situation, at least temporarily. No matter what we do, some amount of short-term economic pain is almost inevitable.

FIDELITY AND HOPE

Of course, what matters most is the long term, and what will determine the long-term fate of modern democracies is not economics but culture. No one will have children to improve the balance sheets of a nation's pension system. Only a new attitude toward procreation and child-rearing can ensure that Western civilization as we know it has a future. That civilization is hopefully not as secular and individualistic as Phillip Longman apparently believes it should be—for if it is, even his pro-child tax proposals are unlikely to inspire the childless to have children or those with one child to have more. But modern civilization is also unlikely to become as religiously serious as George Weigel and others hope. Much as we need a generation moved by Hannah's maternal longings, we cannot rely on her unshakeable faith to fuel it. The Haredi Jews, the Mormons, and the orthodox Catholics who reject modern birth control will continue to reproduce in high numbers. But they will always remain subcultures within modernity; the democratic future cannot rest on their piety alone.

If there is any hope for the modern West, we need a compelling humanistic answer to the question "Why have children?"—and in the plural, not the singular. This answer needs to resonate with those who are open to religious faith but uncertain that God wills them to be fruitful. It needs to appeal to those who appreciate the material benefits of modern life but are not so governed by modern ambitions alone that having more children seems like robbing opportunity from the one or two children they already have. And it needs to demonstrate that in seeking the perfect or perfectly happy child, modern parents would deny their offspring the greatest benefit of all: brothers and sisters with whom to grow up.

The philosopher Gabriel Marcel, in a pair of lectures delivered in Europe in the early 1940s, points us in the right direction. "By this inextricable combination of things from the past and things to come," he said, "the mystery of the family is defined—a mystery in which I am involved from the mere fact that I exist." Even the most modern individual can discover that he is not "endowed with an absolute existence" of his own, but caught up in a web of familial relations whether he likes it or not.[33] By seeing the present in light of the past and the future, the happiness-seeking individual will perhaps make room in his home for a new generation to seek happiness after he is gone. Just as we ourselves are the "incarnation" of past generations, so on our shoulders do the hopes of our ancestors rest; we are links to the future or else we are the end of the past. In our children, the past that we incarnate is potentially preserved or redeemed; in our children, the sweet abundance of the present is appreciated anew; in our children, the future—of our families, our people, and our beliefs—is at least given a chance.

Looking backward, having children is an act of *fidelity* to one's near and distant ancestors, a way to repay the generosity they invested in the children they reared, or to redeem the tribulations of one's own childhood by bestowing generous life on sons and daughters. Looking forward, having children is an act of *hope*, a belief in the possibility of a human future, inhabited not by an immortal self but by novel beings with fresh possibilities. And looking to the present, having children is a way to experience anew the simple delight in *being* that every mature person loses over time, sharing in a life untroubled by the psychological burdens and grave responsibilities of adulthood, filled with wonder at everyday things that we no longer even notice. From one imperfect generation to the next, children renew life in the face of finitude.

Of course, there will always be many reasons to have children— some compelling and some trivial, some selfish and some generous, some morally praiseworthy and some morally troubling. We can have children as future warriors for the state, or to succeed in all the ways that we failed, or to care for us when we are old. We can have children because we believe propagating our genes will improve the race, or to prove that we can accomplish something, or to remedy our loneliness. We can have children to pass on the name of a dead relative, or to save a sibling dying of cancer, or as a gift to the husband or wife whom we love. But the best reasons to have children are never the most specific. For there is a vast gulf, as Marcel put it, between those who "dole out only the smallest possible number of descendants compatible" with their self-regarding purposes and "those who, in a sort of prodigality of their whole being, sow the seed of life without ulterior motive."

There is a great difference between the reluctant parent who subordinates child-rearing to his pre-existing plans and projects, and the would-be parent "who is the bearer of some message which he must communicate, of some flame which he must kindle and pass on."

According to a recent Gallup poll, 70 percent of childless women over 40 regret that they have no children.[34] Such a statistic is reason for great sadness but perhaps also modest hope: sadness for what the childless have never experienced, hope that their sad experience will awaken the generations that follow to embrace the adventure of motherhood and fatherhood, with all the sacrifice and joy involved in raising up one's descendants.

CHAPTER NINE

In Whose Image Shall We Die?

THE STARK REALITY of death persists unaltered through time: we still shudder at the finality of the lifeless corpse; we still shoulder the responsibility of relinquishing the deceased's mortal remains; we still bear the burden of mourning the dead and accepting life without them, often with prayers and rituals of ancient origin. To be human means that death is inevitable; to care for our fellow human beings means that caring for the dead is inescapable. Technological progress and social change do not alter the essence of man's encounter with mortality, with both its mystery and its misery.

Yet in crucial ways, if not the most fundamental ways, how we die in the modern age would be unrecognizable to our forebears. Death often arrives on the doctor's watch, and only after an explicit decision by the dying person's surrogates to forgo additional life-sustaining treatments. In many cases, knowing when or whether death has actually arrived is puzzling: the brain-dead body hooked up to a ventilator is not lifeless—the heart still beats, the chest still moves, the organs still work. But these organs are no longer parts of a living whole. Moreover, once death has occurred, relinquishing the body and mourning the deceased are no longer the only concerns. The irreversibly dead body now has a novel potential—to be used for medical education, or scientific research, or as a supply of organs for patients whose organs are failing.[1]

Of course, death is not the only human problem of bioethical significance. Natality, not mortality, is arguably the source of today's gravest and most novel quandaries—from the prospect of human cloning to the genetic screening of embryos to the return of eugenics in the form of amniocentesis-and-abortion. Yet it is also the case, as explored above, that the very technological civilization that has developed these marvelous new methods of making babies—children for the infertile, children without disorders, children for older women— is also the least interested in procreation, at least by the numbers. Modern, advanced democracies have the lowest birthrates in human history; they are not producing enough children to replace themselves.[2] And it may be that our anti-natalism has much to do with our understanding (or misunderstanding) of our mortal condition. We readily ignore death, making procreation seem less urgent to men and women who think there will always be more time; and we desperately evade death, making procreation seem less important than sustaining the healthy self into the indefinite future. A death-denying civilization is also, it seems, a child-denying civilization.

Moreover, if one considers the most passionate bioethical debates of recent years—embryo research, assisted suicide, and euthanasia— the central question seems to be how to live well with death, or how to care well for those who live on the precipice between life and death. With embryo research, we are forced to ask: Is it better to accept death than to destroy human embryos in an effort to oppose it? Is it better to submit to suffering and surrender to mortality than to use the seeds of the next generation as raw materials in the search for cures? With assisted suicide and euthanasia, we are forced to ask: Is it love or is it torture to keep a profoundly debilitated person alive indefinitely? Is it mercy or is it betrayal to let her die of dehydration by removing her feeding tube? On both fronts, we need to ask: What is the good death, or what is the best death possible for moral beings who must die and who know it?

The moral riddles of our mortality promise to get only more complicated in the years ahead. Modern medicine's capacity to defeat earlier, acute causes of death may also lead, for many of us, to an extended decline into debility and dementia. The heart attack one averts at age sixty-five might lead to a decade of Alzheimer's disease—a gradual erosion of the self much different than the accident victim's overnight debilitation. And those inventions—like dialysis—that were once seen as life-saving miracles are now seen as torturous burdens. Instead of appreciating the thousands of individuals saved each year by transplanting human organs—individuals once destined to die of organ failure—we see the organ waiting list as a "crisis" in need of solution or as an unconscionable failure of public policy.[3] And we wonder: Might it be better to buy organs from the poor or conscript organs from the nearly dead than to accept a death that might be averted? Should we set aside our ethical qualms to save more lives? Whatever one thinks about any particular bioethical issue, the problem of living well with death is integral to them all, even those that seem to center more on natality than mortality. And how one thinks about each particular bioethical issue depends, whether explicitly or unknowingly, on the image of death that we see as best.

To face our most modern dilemmas, therefore, we need to recover some very ancient wisdom. Our culture offers various paradigmatic images of the good death: the remembered death of Jacob; the tranquil death of Socrates; the redeemed death of Christ; the technological opposition to death of Benjamin Franklin; and the crisis of death described in Albert Camus's myth of Sisyphus. These images are the best prism for seeing our past, present, and future bioethical dilemmas for what they truly are. In the age of ventilators and nursing homes and regenerative medicine, we must always ask: In whose image should we die, and in whose image should we live on the way to death? The Bible, as usual, is a good place to start.

THE REMEMBERED DEATH

Jacob is the last of the three great biblical patriarchs, who dies surrounded by his many children. He dies naturally, from illness. He is not killed by an enemy, or lost in a tragic accident, or sentenced to death by a just or unjust court. His death is foreseeable, but there is little reason to believe that he suffered an extended decline. He dies knowing that he is dying, not after years of dementia, when self-awareness of one's impending oblivion is existentially impossible. He faces death frontally, manfully, without illusions. In his final moments, he performs the parting act of instructing his sons in their obligations and prospects.[4] How he does this, and why, is the key to understanding Jacob's image of the good death.[5]

The biblical text begins its account of Jacob's death as follows: "When Jacob was told, 'Your son Joseph has come to see you,' Israel summoned his strength and sat up in bed."[6] Jacob is sick, but he will not address his sons in a sickly posture. He sits up before them; his physical presence embodies both his mortal fragility (in bed) and his paternal majesty (sitting up). Even sitting, he remains the upright master, worthy of reverence, still in command even as his body shuts down. Even as his last speech is a recognition of his own mortal limits, he is never an object of pity in his children's eyes.

But Jacob's death, in the end, is not ultimately about him, but about the way of life that he wishes to persist after he is gone. He recounts how God promised to make him "fertile and numerous, making of [him] a community of peoples."[7] He beholds his grandchildren with special amazement, as the fleshy embodiment of the promise of perpetuation. Then he addresses each one of his children —some with great hope, some with stinging disappointment—for he knows that the fate of his divine purpose rests on their shoulders, a prospect that leaves him to die without the certainty of success, but also without the certainty of failure.[8] He asks his children to remember him, awakening their ancestral piety as the ground for continuing life beyond themselves, in "teeming multitudes upon the earth."[9] He links reverence for the past with hope for the future. He dies, in other words, as the dying father and the dying son. This fidelity in death centers symbolically and ritually on burial—the return of Jacob to the land of his fathers. More importantly, it depends upon the willingness of his children to raise up children of their own, before whom they will one day stand in death, children who will in turn have children of their own, to perpetuate God's holy and hopeful way into the future.

One episode in particular captures this way of dying well. Just before his last speech to all of his sons, when he knows he is dying but before they do, Jacob orders Joseph to put his hand under Jacob's thigh and pledge to bury the dying patriarch in the land of his fathers. The point of the pledge is not simply to satisfy Jacob's self-regarding wishes—to fulfill his advance directive, so to speak. It is also a reminder to Joseph of where he comes from, who he is, and what he must teach his own children.[10] In demanding this oath, Jacob instructs his son never to forget. By demanding that Joseph place his hand under the thigh—in that physical place where the next generation finds its origins—Jacob instructs Joseph of what fidelity really means. One remembers the dead by giving birth to the living; one dies well by giving one's children their final instructions.

Now imagine, instead, if Jacob had put his hand on Joseph's back, on the body of his son, and requested a kidney in the desperate hope to stay alive. Or imagine if he had produced an embryonic clone of himself, nascent flesh of his own flesh, in the hope of manufacturing a life-saving cure. This is strange to imagine, and not merely because of the historical distance between our mythical ancestors and our modern medical practices. Such desperate requests or actions— violating the body of his son, destroying the seeds of the next generation—would pervert Jacob's way of dying well, in which he stands before his sons commanding their fidelity, majestic even in dying. Jacob needs his sons to continue life after he is gone much more than he needs their bodies to extend his life here and now. Yet Jacob's need for his sons—for he is impotent in death without them—never seems needy. Jacob's death makes sense because he stands aside for his children, yet stands above them even in the moment when he needs them most—to remember him, to bury him, and to carry on his sacred purpose. This is the Jewish way of dying well, of living well with death.

It is also why, throughout the story of the patriarchs—indeed, throughout the whole Hebrew Bible—barrenness, not sickness, is the real threat to the good life and the good death; opening the womb is the truest evidence of God's beneficence. Sarah's misery, Rebekah's misery, Rachel's misery, Hannah's misery—all finally answered when God remembers them with a child—is the misery of infertility.[11] Even earlier, in his address to Noah after the flood, God tells man to "be fruitful and multiply"—first as a divine blessing, then as a divine commandment.[12] God seems to realize that the human revolt against children—willful sterility, not un-chosen barrenness—is a permanent human possibility, as men get lost in the ecstasies or the miseries of present life. God seems to know that only man among the animals can choose against the next generation—seeing children as a burden, or seeing life as too burdensome to inflict on another generation.

Yet in our time, in this most death-defying of civilizations, procreation is becoming an afterthought, as noted above. Modern technological societies, infatuated with embryo research and organ transplantation and life-saving cures, are having the fewest children of any societies in human history. Today's potential parents are much less likely to die surrounded by their offspring, or remembered by their children, or sustained by their children's children. From Jacob's perspective, we are dying badly by dying alone, with no sons and daughters to instruct in our final days.

But Jacob's way of dying is also threatened from a different angle in modern societies, also noted above. Today, we are much more likely to die only after an extended demise, after long years of physical and mental decline into dependency and dementia, unable to sit upright before our children in our final days. The very medical triumphs that make long life and prolonged vigor possible for so many can also (if unintentionally) make dying an extended misery. Our medical machinery makes Jacob's version of the good human death ever more unlikely. Even in the eyes of the most devoted children, we risk becoming an object of pity. Or, in our childlessness, we risk becoming a burden on the state. Such circumstances threaten to usher in a new age of euthanasia, both as a way of restoring the social balance between the old and the young and as a way of recovering the tranquil, timely death that most people still want but few people now get.

THE TRANQUIL DEATH

The death of Socrates, as remembered by his student Xenophon, is a very different kind of human death—a noble euthanasia. The philosopher has been convicted by the city for worshipping false gods, or no gods at all, and for corrupting the youth of Athens. Unlike Jacob and his fathers, Socrates does not see his life as the fulfillment of a divine commandment, or his wisdom as dependent on God's revelation. When the oracle of Apollo declares that Socrates is the wisest man alive, he sets out (as Leo Strauss noted) to prove the oracle wrong, only to discover that the oracle is right.[13] Everyone who claims to be wise is actually foolish, believing he knows the truth of ultimate things when in fact he knows nothing. Socrates at least knows that he knows nothing. He also knows when it is time to die, or at least how to die well. He has little interest in admitting guilt, or apologizing, or escaping into exile in order to avoid death. As Xenophon says, Socrates "had come to regard death as for himself preferable to life," and so he accepts his death sentence with a certain tranquility.[14]

Still a great giver of speeches, Socrates also knows that his bodily decline is looming. He seems to abhor the prospect of losing his mental powers, of being alive without the capacity for wisdom, of being an object of pity or contempt to all those who presently admire or fear him. He knows that he cannot be a thriving philosopher forever, and he sees his death, under these circumstances, as a kind of good fortune—one he attributes to a god, but which comes about by his own forced, if willing, human hand. As Socrates says:

It may be . . . that God out of his great kindness is intervening in my behalf to suffer me to close my life in the ripeness of age, and by the gentlest of deaths. For if at this time sentence of death be passed upon me, it is plain I shall be allowed to meet an end which . . . is not only the easiest in itself, but one which will cause the least trouble to one's friends, while engendering the deepest longing for the departed. For of necessity he will only be thought of with regret and longing who leaves nothing behind unseemly or discomfortable to haunt the imagination of those beside him, but, sound of body, and his soul still capable of friendly repose, fades tranquilly away.[15]

Interestingly, the President's Council on Bioethics cites this passage in its 2005 report *Taking Care: Ethical Caregiving in Our Aging Society* as an image of what the good death might look like, in contrast to an extended modern decline into dementia or the awful prospect of dying too soon. But the Council's sympathy for this Socratic death is not without caveats and questions, which it raises in a lengthy footnote:

If we are still sound of body and mind, can we ever really accept death with tranquility? And if we are still a source of happiness to our friends, would they let us "fade away" if they had the power to keep us going? Do human beings deserve the most tranquil death? Or is death, in some ways, the very opposite of tranquility—a nasty robbery of life, to which we can surrender gracefully but never happily? And what is the meaning of the fact that the peaceful death here described (the death of Socrates) is brought about by deliberate—or deliberately imposed—human action (that is, by the drinking of hemlock)? Nevertheless, Xenophon is clearly on to something: a peaceful death, in the right season, is for most of us the best we can humanly hope for.[16]

Unlike Jacob, Socrates dies among students and friends, not among his children. It is his friends' trouble he seeks to avoid; their fond memories ("longing for the departed") he seeks to sustain. That Socrates dies a noble death—a death with dignity—is hard to deny. He stands unflinchingly, almost playfully, before his supposedly pious accusers. And he asks them questions they cannot answer, confronts them with contradictions they cannot ignore, and demonstrates for eternity that independent spirit that belongs to the philosopher alone. Where Jacob accepts a natural death he cannot escape, Socrates accepts an imposed death he might have averted, but averted only by betraying who and what he was. Death thus becomes the vindication of philosophy, of truth opposed to opinion, wise questioning opposed to ignorant certainty, without the wisdom-wrecking decay of the mortal body. It is an upright death, a death that preserves the immortal dignity of the man who died at the summit of his powers and on the eve of his decline. It is also a pleasant death—swift, painless, "easiest in itself." [17]

Yet, for all its renown, the death of Socrates seems less fully human than the death of Jacob, which unites the private drama of father and sons with the public drama of Israel's beginnings as a nation. Jacob's speech, if less grand than the apology of Socrates, seems truer to what it means to live in time, called to a purpose, remembered through the fidelity and perpetuation of one's offspring. And ultimately, the Socratic death embodies a certain ambiguity as both the brave death and the tranquil death. Socrates dies well by accepting death rather than betraying his commitment to truth; yet he also needs death to come sooner rather than later, so that nature does not destroy his nobility as a philosopher by destroying his embodied mind, turning the wisest man into a post-philosophical body. For Socrates, the most pleasant death is, necessarily, the least natural death—the controlled exit, without nature's "unseemly or discomfortable" afflictions "to haunt the imagination of those beside him." [18] But the dignity of this pleasant, unnatural death also seems to require that such a death be un-chosen. The death sentence of Socrates replaces the deathbed of Jacob. The heroic and the tranquil are united in one final sip of hemlock, a poison that the poisoned man sees simultaneously as both an injustice and a gift.

THE REDEEMED DEATH

The death of Jesus is also heroic, but hardly tranquil. Like Socrates, Jesus dies at a time decreed by the civil authorities, not by the entropy of nature. He spends his final hours among his disciples, not his children. Yet unlike Socrates, Jesus dies the most painful death imaginable, an extended public torture, horrifying to those who love him. Like Socrates, his death is imposed upon him by others. But whereas Socrates, at least according to Xenophon, seems to prefer death to life, Jesus dies as an act of submission. And while Jesus is destined in the story to rise again, there is also a way in which, unlike Socrates, Jesus dies before his time; he dies not in the proper season; he dies watched by his mother, not remembered (like Jacob) by children who follow in his footsteps.

Jesus' death is not finally about him, of course, but about the divine purpose he is called to fulfill. Childless, he looks to his disciples to perpetuate his holy way, to preach the gospel, to spread the good news. He dies the paradoxical death—mocked, but dying to redeem the mockers; innocent, but dying to conquer sin; submitting to death, but only so he might conquer death through love. For Jesus, unlike Jacob or Socrates, death has to be a misery, "discomfortable" to those who love him. Even in his innocence, he embodies the fact that death is the wage of sin, a just sentence upon humanity, lifted only by God's grace; death is not the injustice it so often seems. Jesus' death is both in need of redemption (the human Jesus) and the redemptive act itself (the divine Jesus). His death is meant to change death forever, allowing even mere mortals to die wretchedly with the faith that death is not final. As Caitrin Nicol writes: "Jesus' death is a physical display of sin, sister to death since Genesis, and it is sin itself that is most importantly being conquered—not faced, not escaped, not accepted, but actually conquered. As mortality was the consequence of the Fall, the literal undoing of death in the Resurrection of Jesus is there to show that the Fall has really been reversed."[19]

Although he begs in Gethsemane that the cup might pass from his lips, he does not resist it when it comes. When one of his disciples tries to defend him by force, he charges him to put away his sword. Cursed by his enemies, betrayed by his friends, abandoned even for a moment by his Father—"my God, my God, why has thou forsaken me?"—Jesus confronts death and the "power of darkness" with no force save one: the love that triumphs over all.[20]

In Jesus, we see what it means to love in the face of misery, to believe in the face of physical horror, to conquer death by submitting to it. Jacob does not conquer death; he steps aside for the children who will remember him and perpetuate his holy mission. Socrates does not conquer death, but simply removes its sting, by treating it as a great unknown and therefore not a known evil, and by accepting the pleasant exit that is so unnaturally offered to him. But for Jesus, death is understood as a problem; it needs to be conquered, not simply accepted. It stings, yet with God's help, man can love and be loved even in the face of its sting.

One wonders what Socrates would have done if his punishment had been crucifixion, not hemlock: Would he have had Jesus' strength, or might he have sought some escape? And one wonders what Jacob's sons would have done if their father, through nature's malignancies rather than man's, had suffered before them as Jesus did? Could they have stood to witness their father so tortured and still believed in their father's God, and would his desperate state have elicited their pity, or rage, or despair? Unlike Socrates and Jacob, Jesus confronts us with the horror of death endured in all its horribleness: not sought as an exit, yet not escaped at the cost of betraying one's given purpose. In Jesus, we learn what it means to forgo all control and retain all control simultaneously—what it means, passively and actively, to die as an act of surrender.

THE OPPOSED DEATH

Modern man, by contrast, faces death with a different credo: never surrender. For modern man, as for Jesus, death is a problem; mortality is an affront; it needs to be conquered. But the route to conquering death—or trying nobly—is not submission, but cleverness; not faith, but science. The aim is regeneration of the body (a self-made act), not resurrection of the body (a God-dependent act). We have taken a page from old King Asa, who in his illness "sought not to the Lord, but to the physicians."[21] In a wonderful letter to Rev. John Lathrop in 1788, Benjamin Franklin gives voice to this modern sensibility—the thirst to extend life with "useful utensils and instruments." As Franklin writes:

> I have sometimes almost wished it had been my destiny to be born two or three centuries hence. For invention and improvement are prolific, and beget more of their kind. The present progress is rapid. Many of great importance, now unthought of, will before that period be produced; and then I might not only enjoy their advantages, but have my curiosity gratified in knowing what they are to be. I see a little absurdity in what I have just written, but it is to a friend, who will wink and let it pass, while I mention one reason more for such a wish, which is, that, if the art of physic shall be improved in proportion with other arts, we may then be able to avoid diseases, and live as long as the patriarchs in Genesis; to which I suppose we should make little objection.[22]

Franklin, like Socrates, seems to have an equanimity about life and death. He admits that his yearning for an ageless body is a kind of "absurdity," and requests from his friend an understanding "wink." He also acknowledges that such blessings will not arrive in time for him. But his optimism is not simply ironic. He believes that progress will fend off death's many causes, if not defeat death itself, and that science will (almost) restore the lost age of man's timeless innocence, or at least secure a life long enough to satisfy man's many curiosities. He believes that technology is a partial—and perhaps the best available—human answer to death.

Of course, we now live "two or three centuries hence," and we might wonder what Franklin would think about our achievements. The marvels of modern medicine surely outstrip the blunt instruments of his own day. The art of biology holds death at bay; it immunizes us from disease, rescues us from injury, and replaces broken parts with new (or healthy used) ones. But at eighty-four years of age when he died, Franklin's life would still be long by modern standards, though his once uncommon longevity is increasingly routine. Science answers many deadly threats, making the body's longevity less a matter of genetic chance or good fortune, and more a matter of human control. But science has not—cannot—answer death itself. Surely one would be a fool not to see modern medical science as a godlike, perhaps even a God-given, blessing. But one also cannot ignore what Franklin's contemporary Rousseau observed about the effect of technological progress on human desire and human happiness:

> [S]ince men enjoyed very great leisure, they used it to procure many kinds of commodities unknown to their fathers; and that was the first yoke they imposed upon themselves without thinking about it, and the first source of the evils they prepared for their descendants. For, besides their continuing thus to soften body and mind, as these commodities had lost almost all their pleasantness through habit, and as they had at the same time degenerated into true needs, being deprived of them became much more cruel than possessing them was sweet; and people were unhappy to lose them without being happy to possess them.[23]

Of course, the sick still see their cures as blessings; they are still filled with gratitude toward their doctors when they leave the hospital to return again to normal life. But Rousseau is clearly on to something.

His insight is borne out, for example, by the contemporary outcry over the "shortage" of organs for transplant. Once regarded as a miraculous gift for the fortunate few who were able to find a suitable organ, transplantation has, by its own successes, come to be regarded as a necessity. Waiting for an organ has become a novel kind of misery. The miracle of a new organ has become, for those in need, an expectation, such that "being deprived of them [is] much more cruel than possessing them [is] sweet."[24] Organ transplantation is just one example of a more widespread phenomenon. All too often, our modern medical technologies are transformed in our eyes from achievements to failures, precisely because they cannot fend off death itself or reverse the ravages of natural time that they temporarily hold at bay. The blessing of dialysis becomes a curse in just a few decades. The diseases of old age come to be seen as epidemics, turning life itself into a war against disease—a permanent, restless march for a cure.

THE CRISIS OF DEATH

Perhaps this is why Albert Camus's modern hero is the embattled doctor in plague-time, with the distance between plague-time and normal-time blurred by the omnipresence and omnipotence of death.[25] In Camus's myth of Sisyphus, Franklin's yearning for indefinite life becomes a rage against death. Death becomes a crisis, not just a problem. Perhaps the difference is that Sisyphus knows death firsthand, in all its wretched blankness. He dies and then returns; his passion for life comes from knowing the alternative of nothingness.

> But when he had seen again the face of this world, enjoyed water and sun, warm stones and the sea, he no longer wanted to go back to the infernal darkness. Recalls, signs of anger, warnings were of no avail. Many years more he lived facing the curve of the gulf, the sparkling sea, and the smiles of earth. A decree of the gods was necessary. Mercury came and seized the impudent man by the collar and, snatching him from his joys,

led him forcibly back to the underworld, where his rock was
ready for him.[26]

Whereas Socrates sees his tranquil death as a divine gift, Sisyphus sees
death as a divine theft, to be opposed (futilely) with all his mortal might.

In Camus, Franklin's desire for life is taken to passionate extremes.
The passion of Sisyphus is more like the passion of Christ, but without the redemptive victory. Instead of the long hours of crucifixion
followed by the eternity of resurrection, Sisyphus faces the permanent
recurrence of pushing a rock up a hill, never reaching the top, always
rolling back down to the underworld, never fully rising again. For Sisyphus, opposition to death is everything, but success is impossible. There
is, at most, a brief moment of existential satisfaction, when the rock lies
still near the top, before beginning again its eternal slide to nothingness.

In Sisyphus, Camus believes he has found an answer to the modern crisis of death: heroic revolt, ending in knowing acceptance of
futility, a knowledge that makes man superior to the absurdity of his
fate. "The lucidity that was to constitute his torture at the same time
crowns his victory. There is no fate that cannot be surmounted by
scorn."[27] To some, perhaps, such scornful stoicism is satisfying, but
for most people it is not. They prefer to look away from death until it
stares them in the face; and when it does, they seek Franklin's help,
hoping the cleverness of science can triumph one more time over the
oblivion that terrifies them.

Modern science thus takes up the mantle of death-as-crisis; the
ethic of triage makes ordinary morality seem absurd in the face of
death's permanent absurdity. This point has been described beautifully by Yuval Levin, reflecting upon the deeper meaning of our current debates over embryo research:

> [I]f the fight against disease writ large—indeed the fight against
> natural death—is an emergency, and if . . . it is a struggle we
> can never expect fully to win, then we must always live in a state
> of emergency. We should be always in a crisis mode, always

pulling out all stops, always suspending the rules for the sake of a critical goal. And that means, in effect, that there should be no stops and no rules; only crisis management and triage. . . . But if life is always at risk and we are always in crisis, then we must always do things that moral contemplation would suggest are wrong. If we are always in a mode of triage, then we must always choose the strong over the weak because they have a better chance at benefiting from our help.[28]

The trouble is that in this war against disease and death, we risk undermining the ideals we profess to hold most dear, beginning with the ideal of human equality. We are tempted to treat the most vulnerable as tools to sustain us in the struggle against death. And when this fight must end inevitably in the defeat we cannot avert, we are tempted to violate equality yet again, by treating the old and debilitated (including the future self) as "lives unworthy of life," as unsightly evidence of our failure. Without Jacob's remembering children, without Jesus' saving faith, without Franklin's triumphant method, we are left in the condition of Sisyphus: faced with the crisis of death we cannot conquer, trapped in a mortal condition we seem ill-equipped to endure. In modern times, the hemlock of Socrates seems ever more appealing, requested in desperation rather than accepted in nobility. In an aging society, in which the elderly come to seem and come to feel like paralyzing burdens, the seduction of euthanasia may be too strong to resist.

DEATH AND BIOETHICS

In the late 1960s and early 1970s, Paul Ramsey (one of the founders of modern American bioethics) wrote extensively about the ethics of end-of-life care. Ramsey's central worry was the technological dehumanization of death.[29] He feared the transformation of dying persons into objects, with a humane exit made impossible by the unyielding machinations of the modern hospital, with its modern ethic of "never surrender."[30] A decade or so later, however, Ramsey was far more worried about the opposite problem: terminating life-sustaining treatment too early; treating the debilitated as "better off dead"; defining as "futile" those who could never be restored to normal but whose lives were hardly over.[31] Ramsey did not, in that period, undergo a philosophical transformation; rather, the facts on the ground changed, and so did his bioethical concerns and priorities.[32]

The development of Ramsey's work reminds us that we are all creatures of time and place, and that the bioethical concerns of the present and future require seeing man's permanent problems in light of changing technological and social conditions, and in light of those philosophical orthodoxies that reign supreme in both our bioethics institutions and the culture as a whole. Yet the deepest problems of the age are always connected to the deepest problems of every age; to face the perils of the moment we need, or should want, to draw upon the wisdom of those ancient giants in whose image our lives are rightly understood and judged.

In modern America, Franklin's technological spirit—the will to oppose death through science—hardly needs additional support. The National Institutes of Health budget has risen dramatically in the past decade, no matter how large the federal deficit or how perilous the condition of our entitlement programs for the elderly.[33] Likewise, in our society, the spirit of liberation hardly needs additional moral support. We are liberated from unwanted conceptions; liberated from unwanted births; liberated from the responsibility of rearing disabled children; liberated from the responsibility of providing economically for our elders; liberated to nip and tuck our given bodies in the name of pursuing happiness. Too little autonomy, like too little enthusiasm for scientific progress, hardly seems like our most pressing bioethical problem.

The real challenge upon which the future of American bioethics will turn is learning how to live and die without violating human dignity in the name of medical research, and learning how to step aside for the next generation without treating the debilitated elderly with a fiscally responsible inhumanity. We need to recover, as best we can,

Jacob's way of dying well: *naturally*, without endless machines or swift poisons; *surrounded by his children*, all assembled at the bedside and prepared to honor the dying patriarch by having children of their own; *frail but upright*, with neither the delusion of endless life nor the burdens of an extended decline into dementia.

This image of human dignity in life and in death hardly translates into a ready-made recipe for dealing with every current or foreseeable bioethical dilemma. But it might shape our moral intuitions and cultural aspirations, from which our law and our policies ultimately derive. That is, we might become more willing to let our cherished elders die, within an ethical and legal framework that prohibits euthanasia and assisted suicide; more open to the responsibility to have and raise multiple children, rather than seeking the freedom that childlessness uniquely offers; more welcoming of children unconditionally, rather than subjecting them to the inegalitarian litmus tests of modern genetic screening; and more devoted to the unique human excellence required to be loving caregivers and noble patients, who forgo their plans and accept their mortality rather than mistreat the vulnerable or betray their fellow men.

Looking around, it is easy to be disheartened about bioethics and the human future: birthrates are falling, the incidence of dementia is rising, genetic screening and abortion and embryo destruction are becoming more commonplace. But so long as we remain open to persuasion, open to the recovery of forgotten images of man, our present errors will always be amenable to future reformations. And as we look around and ahead, we should never forget that every age is twisted in its own unique way, stained by errors, getting worse even as it gets better. The only ineradicable error is believing that all the problems of human life can be solved once and for all. Short of that, we will muddle through in bioethics as we do in every realm of human life where the meaning of our humanity is on trial—with examples of excellence and depravity, but most of us stuck in that imperfect in-between, neither beasts nor gods but men with birthmarks.

ACKNOWLEDGEMENTS

In writing a book such as this one, an author incurs many debts, both intellectual and personal. Thanks are owed first to my parents, to whom this book is dedicated. Only in becoming a father (and the husband of a mother) does a child come to understand the gratitude he owes his parents, especially when they devote their lives first and foremost to the flourishing of their four sons. To them, the best thanks I can give is to try to raise my own children with the same full measure of devotion.

Daniel Kincaid and Christopher Watts, my teacher and roommate at Groton School, have been friends in every way for more than fifteen years and counting. Robert Jackall and James Nolan, my teachers at Williams College, played a crucial role in my intellectual formation, opening up for me the great social, philosophical, and moral questions of modernity. Adam Wolfson, my mentor at *The Public Interest*, taught me how to be an editor. He is a master wordsmith, a superb mind, and a great friend. Robby George, a collaborator on many ventures large and small, has been a constant supporter of my work and a model of how to combine intellectual seriousness, moral purpose, and the kind of human dignity one learns growing up in West Virginia.

To the Kristols—Irving and Bill—I owe a great deal, both personally and intellectually. Irving Kristol gave me my first job as an assistant editor at *The Public Interest*. He also gave me, with his essays, a philosophic and political education, connecting the deepest matters of the soul to the most pressing matters of public life. Reading these essays remains one of my great intellectual pleasures, though it is always more than a mere pleasure; and when I find myself unable to write on a morning when writing is required, I grab one of the Kristol collections off the shelf, pick something at random, and read. Bill Kristol has been a mentor, collaborator, and friend for the past many years. Many of the ideas in this book took shape in conversations with Bill, whose capacity to move seamlessly and seriously from topic to topic is both rare and remarkable. He is a singular figure in American life—a deep student of political thought, a force in American politics, and the best political commentator in the business.

I also owe a great deal to Gilbert Meilaender. This debt should be obvious to anyone who knows his work and reads this book: from his writings, always so deep, always so clear, always so hard-headed and so elevating at the same time, I have learned more than I can properly acknowledge in the formality of footnotes. But what I most treasure about my friendship with Gil are the lunches—nearly every month for nearly five years, conversations about the ethics of everyday life, where for me Gil was and remains a model of how to balance the life of the mind and, well, life itself.

Some of the chapters in this book first took shape on the pages of *Commentary* ("The Human Difference," "Why Have Children?") and *First Things* ("The Ends of Science," "A Jewish-Catholic Bioethics?"), and I want to thank those publications for permitting me to use them here. More importantly, I want to thank Neal Kozodoy and Jody Bottum. Neal is a master editor and great friend. His towering intellect is matched only by his unending generosity. Jody has one of the liveliest minds of his generation; working with him always opens new intellectual doors in unexpected ways.

Without the Ethics and Public Policy Center, my intellectual home for more than five years, this book would never have been written. For this, I owe a great deal to Hillel Fradkin, who had both the intellectual imagination and the nerve to allow a very young man to launch a new program in bioethics and a new magazine on the moral challenges of modern science and technology. And I owe a great deal to Ed Whelan, who has made EPPC an intellectual force in Washington and a true community of thinkers and friends. I thank him for his leadership, his friendship, and his unyielding character. Caitrin Nicol and Mary Rose

Rybak both worked tireless hours moving this book from manuscript to completion. One could not ask for more devoted colleagues; one could not imagine two finer people. I also want to thank Roger Kimball and the whole staff at Encounter Books for making this project possible, and Roger Hertog for giving me a new home and new purpose at the Tikvah Fund.

Beyond my family, my deepest debts, intellectually and humanly, are to Leon and Amy Kass, Yuval Levin, and Adam Keiper. To have such teachers, such friends, such colleagues is more than any human being can ask for, and more than this human being can ever hope to repay. How many times have I turned to Leon and Amy for help and advice on the gravest human matters; how many intellectual doors have opened by studying Leon's writings, or conversing at the Kass dinner table, or working together on the many reports of the President's Council on Bioethics. What it means to be father and son, husband and scholar, human being and Jew—Leon and Amy have helped show me the way. In acknowledging this debt, words simply fail.

In Yuval and Adam, I have the finest friends any man could ever ask for. Working with them for the past five years—running a presidential commission, starting a magazine, helping to build a think-tank—has been the greatest privilege of my life. My admiration for each of them, as great minds and great souls, is deeper than I can say. And my gratitude for their many acts of devotion, and for our many conversations, is greater than I can properly express here.

In my beloved wife Stephanie and my darling children Gavriella and Nethaniel, I have found the love and the hope that give man a glimpse of what lies beyond progress. In counting my blessings, they are for me, always, the first and the greatest.

ERIC COHEN
March 2008

NOTES

CHAPTER ONE · THE SPIRIT OF MODERN SCIENCE

1 John Bohannon, "Disasters: Searching for Lessons From a Bad Year," *Science* 310 (December, 23 2005): 1883.

2 See, e.g., Michael Gazzaniga, *Personal Statement in Reproduction and Responsibility* (Washington, DC: President's Council on Bioethics, 2004), 238; and Michael Gazzaniga, *The Ethical Brain: The Science of Our Moral Dilemmas* (New York: HarperCollins, 2005), 11. For a critique of this argument, see Editorial, "Morals and the Mind: Michael Gazzaniga's Ethical Brain," *The New Atlantis*, no. 11 (Winter 2006) 121–125.

3 Edward O. Wilson, "Let's accept the fault line between faith and science," *USA Today*, 15 January 2006.

4 Richard Dawkins, *A Devil's Chaplain* (Boston: Mariner Books, 2004), 128, 13. See also Richard Brodie, *Virus of the Mind: The New Science of the Meme* (Seattle: Integral Press, 1996).

5 Psalm 27:1 (KJV).

6 Francis Bacon, *The Great Instauration* in *The English Philosophers from Bacon to Mill* (New York: Random House, 1939), 13, 18. Originally published in 1620.

7 Francis Bacon, "New Atlantis" in *The Great Books of the Western World*, vol. 30 (Chicago: University of Chicago Press, 1952), 211. Originally published in 1627.

8 Alfred North Whitehead, *Science and the Modern World* (New York: Free Press, 1997), 10. Originally published in 1925.

9 Charles Darwin, *The Origin of Species* in *The Great Book of the Western World*, vol. 49 (Chicago: University of Chicago Press, 1952), 243. Originally published in 1859.

10 *Ibid.*

11 Francis Bacon, *Novum Organum* in *The English Philosophers from Bacon to Mill* (New York: Random House, 1939), 65. Originally published in 1620.

12 Marie Jean Antoine Nicolas de Caritat, marquis de Condorcet, *Sketch for a Historical Picture of the Progress of the Human Mind* (New York: Noonday Press, 1955), 124. Originally published in 1795.

13 *Ibid.*, 4.

14 Max Weber, "Science as a Vocation," from *Max Weber: Essays in Sociology* (New York: Routledge, 1998), 139–140. This essay was originally delivered as a lecture at Munich University in 1918.

15 *Ibid.,* 156.
16 José Manuel Rodríguez Delgado, *Physical Control of the Mind: Toward a Psychocivilized Society* (New York: Harper & Row, 1971).
17 Hans Jonas, *The Phenomenon of Life: Toward a Philosophical Biology* (Evanston: Northwestern University Press, 2001), 195. Originally published in 1966.
18 *A Devil's Chaplain,* 246–247.
19 Psalm 23:4.
20 *A Devil's Chaplain,* 13.
21 For a discussion of this theme, see Gilbert Meilaender, "Between Beasts and God," *First Things,* no. 119 (January 2002), 23–29.
22 Richard Dawkins, *The God Delusion* (Boston: Houghton Mifflin, 2006), and Daniel C. Dennett, *Breaking the Spell: Religion as a Natural Phenomenon* (New York: Viking Adult, 2006).

CHAPTER TWO · THE HUMAN DIFFERENCE

1 The State of the Union Address by the President of the United States, 109th Cong., 2d sess., Congressional Record 31 (January 31, 2006): H 15–19. (Available online at http://www.whitehouse.gov/stateoftheunion/2006/.)
2 Ying Chen *et al.,* "Embryonic stem cells generated by nuclear transfer of human somatic nuclei into rabbit oocytes," *Cell Research* 13, no. 4 (2003): 251–263. Michael Hopkin, "Britain gets hybrid embryo go-ahead," *Nature,* September 5, 2007. Jamie Shreeve, "The Other Stem Cell Debate," *New York Times,* April 10, 2005. Brenda Ogle *et al.,* "Spontaneous fusion of cells between species yields transdifferentiation and retroviral transfer in vivo," *Journal of the Federation of American Societies for Experimental Biology* 18 (2004): 548–550.
3 H. G. Wells, *The Island of Dr. Moreau* (New York: Modern Library, 2002), 56. Originally published in 1896.
4 Committee on Guidelines for Human Embryonic Stem Cell Research, National Academies, *Guidelines for Human Embryonic Stem Cell Research.* Washington, DC, 2005, 50.
5 This is captured beautifully in the Yom Kippur liturgy: "Man's superiority to the beast is an illusion. All life is fleeting breath. But You distinguish man from the start, deeming him worthy to stand in Your Presence." Rabbi Jules Harlow, ed., *Mahzor for Rosh Hashanah and Yom Kippur: A Prayer Book for the Days of Awe* (New York: The Rabbinical Assembly), 743.
6 Harold Varmus, "Is Science Under Siege?" (lecture, American Academy of Arts and Sciences, New York, NY, November 16, 2005).
7 *Guidelines for Human Embryonic Stem Cell Research,* 50.
8 William Dembski, (lecture, Fellowship Baptist Church, Waco, TX, March 7, 2004).
9 Psalm 19:1.

10 See, for example, Michael J. Behe, *Darwin's Black Box* (New York: Free Press, 1996).

11 The meaning of the human difference is usefully explored by Mortimer Adler in his book by that title. Mortimer Adler, *The Difference of Man and the Difference It Makes* (New York: Fordham University Press, 1993).

12 Hans Jonas, "Tool, Image, and Grave: On What is Beyond the Animal in Man," ed. Lawrence Vogel, *Mortality and Morality: A Search for the Good after Auschwitz* (Chicago: Northwestern University Press, 1996), 75.

13 *Ibid.*, 76.

14 *Ibid.*, 78.

15 *Ibid.*, 79.

16 *Ibid.*, 83. Humans alone pay recognition to their dead, with perhaps the exception of elephants. Karen McComb *et al.*, "African elephants show high levels of interest in the skulls and ivory of their own species," *Biology Letters* 2, no. 1 (March 22, 2006), 26.

17 "Tool, Image, and Grave," 84.

18 Charles Darwin, *The Autobiography of Charles Darwin 1809-1882* (New York: W. W. Norton & Company, 1993), 232–234. Originally published in 1887.

19 *Ibid.*

20 Aeschylus, *Prometheus Bound*, ed. Paul Elmer More (New York: Houghton Mifflin, 1899), 60.

21 Hannah Arendt, "The Conquest of Space and the Stature of Man," in *Between Past and Future* (New York: Penguin, 1968), 277.

22 This is beautifully explored in Leon R. Kass, *The Hungry Soul: Eating and the Perfecting of Our Nature* (Chicago: University of Chicago Press, 1999).

23 Psalm 23:2; Genesis 6:13 (KJV).

CHAPTER THREE · BIOETHICS IN WARTIME

1 C. S. Lewis, "Learning in War-Time," in *The Weight of Glory* (New York: Harper Collins, 2001), 47.

2 *Ibid.*, 49.

3 President George W. Bush, Address, "Remarks to the Woodrow Wilson International Center for Scholars," *Weekly Compilation of Presidential Documents* (December 14, 2005): 1856. Also available online at http://www.whitehouse.gov/news/releases/2005/12/20051214-1.html.

4 "Learning in War-Time," 53.

5 *Ibid.*, 62.

6 Leon Kass addressed a similar theme in his acceptance speech for the Bradley Prize in 2004.

7 Ian Wilmut, *et al.*, "Viable offspring derived from fetal and adult mammalian cells," *Nature* 385, no. 6619 (1997): 810–813. James Thompson, *et al.* "Embryonic

Stem Cell Lines Derived from Human Blastocysts," *Science* 282, no. 5391 (1998): 1145–1147. "President Clinton Announces the Completion of the First Survey of the Entire Human Genome," press release from the office of the press secretary, the White House (June 25, 2000); available at the William J. Clinton Presidential Library, National Archives. Shaoni Bhattacharya, "Controversial three-parent pregnancy revealed," *New Scientist* (October 14, 2003). Ying Chen *et al.*, "Embryonic stem cells generated by nuclear transfer of human somatic nuclei into rabbit oocytes," *Cell Research* 13, no. 4 (2003): 251–263. Rick Weiss, *Washington Post*, "Scientists Produce Human Embryos of Mixed Gender," July 3, 2003.

8 The President's Council on Bioethics, *Beyond Therapy: Biotechnology and the Pursuit of Happiness*, October 15, 2003.

9 Shaoni Bhattacharya, "Controversial three-parent pregnancy revealed," *New Scientist* (October 14, 2003).

10 Christine Rosen, "Why Not Artificial Wombs?," *The New Atlantis*, no. 3, 67–76. Ying Chen *et al.*, "Embryonic stem cells generated by nuclear transfer of human somatic nuclei into rabbit oocytes," *Cell Research* 13, no. 4 (2003): 251–263.

11 Job 34:5.

12 "Learning in War-Time," 49.

13 Jonas, *The Phenomenon of Life*, 5.

CHAPTER FOUR · THE EMBRYO QUESTION

1 Cited in Leon R. Kass, "The Meaning of Life—In the Laboratory," *The Public Interest*, no. 146 (Winter 2002), 45–6.

2 Robert Edwards and Patrick Steptoe, *A Matter of Life: The Story of a Medical Breakthrough* (New York: William Morrow, 1980), 186–7.

3 See Genesis 18:10–15 and 21:1–7.

4 See Peter Singer, *Practical Ethics* (Cambridge, England: Cambridge University Press, 1993), 173.

5 Harvey C. Mansfield, Jr., "Returning to the Founders: The Debate on the Constitution," *The New Criterion* 12 (September 1993).

6 Yuval Levin, "The Paradox of Conservative Bioethics," *The New Atlantis*, no. 1 (Spring 2003), 62.

7 Louisiana Revised Statutes 9:126 (2007).

8 See Sebastiaan Mastenbroek *et al.*, "In Vitro Fertilization with Preimplantation Genetic Screening," *New England Journal of Medicine* 357, no. 1 (July 5, 2007), 61–3.

9 See Rebecca Dresser, "At Law: Human Cloning and the FDA," *Hastings Center Report* 33, no. 3 (May/June 2003), 7–8.

CHAPTER FIVE · OUR GENETIC CONDITION

1 James D. Watson, "Moving Toward the Clonal Man," *Atlantic Monthly*, May 1971. (This article is a slightly modified version of Watson's congressional testimony.)

2 Gerald Schatten, testimony before the President's Council on Bioethics, December 13, 2002, Washington, D.C. (http://www.bioethics.gov/transcripts/deco2/session6.html).

3 See Anna Salleh, "Cloning humans, primates may be impossible," *ABC Science Online*, April, 11 2003 (http://www.abc.net.au/science/news/stories/s830381.htm).

4 J. A. Byrne *et al.*, "Primate embryonic stem cells by somatic cell nuclear transfer," *Nature* 450, (November 22, 2007), 497–502.

5 Lenny Moss, *What Genes Can't Do* (Cambridge: M.I.T. Press, 2003), 185. Quoted in Steve Talbott, "Logic, DNA, and Poetry," *The New Atlantis*, no. 8 (Spring 2005), 66.

6 Leon Kass, *Life, Liberty, and the Defense of Dignity* (San Francisco: Encounter, 2002), 293. As great as their scientific achievement surely was, Watson and Crick's famous declaration that they had discovered "the secret of life" was palpably naïve. [James D. Watson, *The Double Helix* (New York: Touchstone, 2001), 197. Originally published in 1968.]

7 René Descartes, *Discourse on Method*, in *The Philosophical Works of Descartes*, vol. I, ed. Elizabeth S. Haldane and G. R. T. Ross (Cambridge, Eng.: Cambridge University Press, 1931), 119–120. Originally published in 1637.

8 This problem is explored in great detail in *Beyond Therapy*, a report produced by the President's Council on Bioethics. I was privileged to help draft this section of the Council report in my capacity as a senior consultant. See "Chapter 2: Better Children" and "Chapter 3: Superior Performance" in *Beyond Therapy* (New York: HarperCollins, 2003), 27–158. Also available online at www.bioethics.gov.

9 Psalm 90:12 (ASV), as quoted in Hans Jonas, "The Burden and Blessing of Mortality," *Mortality and Morality: A Search for the Good after Auschwitz*, ed. Lawrence Vogel (Evanston: Northwestern University Press, 1996), 87. This essay was first presented to the Royal Palace Foundation in Amsterdam on March 19, 1991.

CHAPTER SIX · THE COMMERCE OF THE BODY

1 Max Weber, *The Protestant Ethic and the Spirit of Capitalism* (New York: Routledge Classics, 2001), 41. Originally published in 1904–1905.

2 John Wesley, "Thoughts Upon Methodism," *The Works of the Reverend John Wesley*, vol. XV (London: Printed at the Conference-Office, 14, City-Road, by Thomas Cordeux, 1812), 332–3. Originally published in 1786. Weber quotes part of this passage in *The Protestant Ethic*, 118–9. Punctuation has been modernized for clarity.

3 Weber 119.

4 Voltaire, "The Worldling," *The Works of Voltaire: A Contemporary Version*, trans. William F. Fleming, vol. XXXVI (Akron, OH: The Werner Company, 1901), 84, and "A Philosophical Dictionary," vol. XII, 295. Originally published in 1736 and 1764.

5 Jerry Z. Muller, *The Mind and the Market: Capitalism in Western Thought* (New York: Knopf, 2002), 30.

6 Voltaire, vol. XXXVI, 85.

7 See Irving Kristol, "Adam Smith and the Spirit of Capitalism," *Neoconservatism: The Autobiography of an Idea* (New York: Free Press, 1995), 258–299.

8 Adam Smith, *An Inquiry into the Nature and Causes of the Wealth of Nations* (Washington: Regnery, 1998), 788.

9 Irving Kristol, "The Cultural Revolution and the Capitalist Future," *Neoconservatism: The Autobiography of an Idea* (New York: Free Press, 1995), 134–135. Reprinted from "The Capitalist Future," the 1991 Francis Boyer Lecture at the American Enterprise Institute in Washington, D.C., December 4, 1991 (http://www.aei.org/publications/pubID.1674,filter.all/pub_detail.asp).

10 This phrase has been used (in a different fashion) especially in connection with theologians Shailer Mathews and Karl Barth. See, for example, William D. Lindsey, Shailer Mathew's *Lives of Jesus: The Search for a Theological Foundation for the Social Gospel* (Albany: SUNY, 1997); and Ingolf Dalfert, "Karl Barth's Eschatological Realism," in Stephen W. Sykes, ed., *Karl Barth: Centenary Essays* (Cambridge: Cambridge University Press, 1989), 22.

11 See Herbert Marcuse, *One-Dimensional Man* (Boston: Beacon, 1964).

12 Peter Schjeldahl, "England Swings: Tate Modern, the Saatchi Gallery, and competing views of the contemporary," *The New Yorker*, March 1, 2004.

13 Sally Wadyka, "For Women Worried About Fertility, Egg Bank is a New Option," *New York Times*, September 21, 2004; Christian Ianzito, "Putting Your Eggs in a Different Basket," *Washington Post*, September 21, 2004.

14 See David Amsden, "Mike Ditka wants to help you score," *Salon*, March 19, 2004.

15 Leon R. Kass, *The Beginning of Wisdom: Reading Genesis* (Chicago: University of Chicago Press, 2006), 209.

16 See, e.g., Hans Scholer *et al.*, "Derivation of Oocytes from Mouse Embryonic Stem Cells," *Science* 300, no. 5623 (May 23, 2003), 1251–1256.

CHAPTER SEVEN · A JEWISH–CATHOLIC BIOETHICS?

1 William Saletan, "Oy Vitae: Jews vs. Catholics in the Stem Cell Debate," *Slate*, March 12, 2005, http://slate.com/id/2114733/.

2 Elliot N. Dorff, "Embryonic Stem Cell Research: The Jewish Perspective," *United Synagogue Review* (Spring 2002).

3 *Ibid.*

4 *Ibid.*

5 *Zohar*, Shemot 3b.

6 As quoted in J. David Bleich, *Contemporary Halakhic Problems*, vol 1 (New York: KTAV Publishing House, Inc./Yeshiva University Press, 1977), 331.

7 As quoted in *Ibid.*, 341–342.

8 As quoted in J. David Bleich, "Abortion in Halakhic Literature," in *Jewish Bioethics* ed. Fred Rosner and J. David Bleich (Hoboken: KTAV Publishing House, Inc., 2000), 157.

9 Exodus 21:22.

10 This analogy appears in *Human Cloning and Human Dignity: An Ethical Inquiry*, a report produced by the President's Council on Bioethics. I was privileged to help draft this report in my capacity as a senior consultant to the Council. See "Chapter 6: Ethics of Cloning-for-Biomedical-Research," 156. Also available online at www.bioethics.gov.

11 Ecclesiastes 11:5 (Jewish Publication Society, Tanakh).

12 J. David Bleich, "Stem Cell Research," *Tradition*, vol. 36 no. 2 (Summer 2002), 77–78.

13 Leon R. Kass, "Human Frailty and Human Dignity," *The New Atlantis*, no. 7 (Fall 2004/Winter 2005), 110.

14 Babylonian Talmud: *Tractate Yebamoth*, Folio 64a.

15 I Samuel 1:11 (JPS).

16 Yitzchok Breitowitz, "What's So Bad About Human Cloning?," *Kennedy Institute of Ethics Journal* 12, no. 4 (December 2002), 331.

17 See, for example, Leon R. Kass, "Making babies—the new biology and the 'old' morality," *The Public Interest*, no. 26 (Winter 1972), 18–56; and Leon R. Kass, "Preventing a Brave New World: Why We Should Ban Human Cloning Now," *The New Republic*, (May 21, 2001), 30–39.

18 A rich account of the Catholic understanding of the body is offered in John Paul II, *The Theology of the Body: Human Love in the Divine Plan* (Boston: Pauline, 1997).

CHAPTER EIGHT · WHY HAVE CHILDREN?

1 Phillip Longman, *The Empty Cradle: How Falling Birthrates Threaten World Prosperity and What to Do About It* (New York: Basic Books, 2004). Ben J. Wattenberg, *Fewer: How the New Demography of Depopulation Will Shape Our Future* (Chicago: Ivan R. Dee, 2005).

2 Vladimir Putin, "Annual Address to the Federal Assembly," delivered at Marble Hall, the Kremlin, Moscow, May 10, 2006 (http://www.kremlin.ru/eng/speeches/2006/05/10/1823_type70029type82912_105566.shtml); Pope Benedict XVI, "Letter to the Participants in the Twelfth Plenary Session of the Pontifical Academy of Social Sciences," the Vatican, April 27, 2006 (http://www.vatican.va/

holy_father/benedict_xvi/letters/2006/documents/hf_ben-xvi_let_
20060427_social-sciences_en.html).

3 Wattenberg, *Fewer*, 24–5.

4 Central Intelligence Agency, *The World Factbook*, 20 September 2007 (https://
www.cia.gov/library/publications/the-world-factbook/geos/ja.html#People).

5 Central Intelligence Agency, *The World Factbook*, 20 September 2007 (https://
www.cia.gov/library/publications/the-world-factbook/geos/xx.html#People).

6 Population Reference Bureau, *2006 World Population Sheet* (August 2006),
8–10 (http://www.prb.org/pdfo6/06WorldDataSheet.pdf).

7 Longman, *Empty Cradle*, 8.

8 RAND Corporation, *Low Fertility and Population Ageing: Courses, Consequences,
and Policy Options* (European Commission, 2004).

9 Phillip Longman, "The Return of Patriarchy," *Foreign Policy*, no. 153 (March/
April 2006).

10 George Weigel, *The Cube and the Cathedral* (New York: Basic Books, 2005), and
"Europe's Two Culture Wars," Commentary 121, no. 5 (May 2006).

11 Mark Steyn, "It's the demography, stupid," *The New Criterion* 24, no. 1 (January
2006), 10. See also Steyn's book, *America Alone: The End of the World as We Know
It* (Washington: Regnery, 2006), 41.

12 As we evaluate the current demographic picture, it is important to remember
that there is no single explanation for fertility rates from place to place, no necessary relationship between economic conditions and birthrates, no simple cor-
relation between religiosity and fecundity. Subcultures within the wealthiest
nations—like the Haredi Jews of New York or the Mormons of Utah—have fertility rates that are among the highest in the world, even as their next-door neighbors have fertility rates that are among the lowest. [Eli Berman, "Sect, Subsidy,
and Sacrifice: An Economist's View of Ultra-Orthodox Jews," *The Quarterly Journal of Economics* 115, no. 3 (August 2000), 905–953. Kevin McQuillan, "When
Does Religion Influence Fertility?", *Population and Development Review* 30, no. 1
(March 2004), 25–56.] Among the modern democratic nations, the richest—
like the United States and France—often have comparatively higher fertility
rates, while the less wealthy—like Poland and Hungary—have comparatively
lower fertility rates. (See CIA's *World Fact Book*, https://www.cia.gov/library/
publications/the-world-factbook/) Among the Islamic countries, Saudi Arabia
and Afghanistan have among the highest fertility rates in the world, while Iran
has one of the lowest. Nor is there a necessary relationship between fertility and
nationalism: In the 1980s, the Iranian government commanded its citizens to
produce more "soldiers for Islam" by increasing fertility. Today, the Iranian government punishes families that have three children or more. [Mohammed Jalal
Abassi-Shavazi, "Recent Changes and the Future of Fertility in Iran," paper prepared for UNPD Expert Group Meeting on Completing the Fertility Transition,

New York, March 11–14, 2002 (http://www.un.org/esa/population/publications/completingfertility/2RevisedABBASIpaper.PDF).]

13 Genesis 1:28 (KJV).

14 Polybius, *The Histories*, Book XXXVI (circa 150 b.c.). As quoted in Steyn, *America Alone*, 41.

15 Hesiod, *Theogony, Works and Days* (New York: Oxford University Press, 1999), 48. Original circa 700 B.C.

16 Aristotle, *Politics*, trans. Ernest Barker (Oxford, Eng.: Clarendon Press, 1948), 327.

17 "... *portentosos fetus extinguimus, liberos quoque, si debiles monstrosique editi sunt, mergimus...*" Seneca, *De Ira* (1.15).

18 Angus McLaren, *A History of Contraception: From Antiquity to the Present Day* (Oxford: Blackwell Publishing, 1991).

19 Longman, "The Return of Patriarchy."

20 Sextus Propertius, *Poems*, trans. Constance Carrier (Bloomington: Indiana University Press, 1963), 67.

21 See I Samuel 1.

22 Plutarch, *Plutarch on Sparta*, trans. Richard J. A. Talbert (New York: Penguin Classics, 1988), 159–160.

23 Neil Gilbert, "What Do Women Really Want?," *The Public Interest*, no. 158 (Winter 2005), 21–38.

24 Elaine Tyler May, *Barren in the Promised Land* (New York: HarperCollins, 1995).

25 Roslyn Smith, "Pity the Childless Couple," *American Mercury*, no. 84 (1957), 76–8.

26 Ellen Peck, *The Baby Trap* (New York: David McKay, 1971).

27 Linda Hirshman, "Homeward Bound," *The American Prospect* 16, no. 12 (December 2005). See also Hirshman's *Get to Work: A Manifesto for Women of the World* (New York: Viking, 2006).

28 Paul R. Ehrlich, *The Population Bomb* (Cutchogue, NY: Buccaneer, 1995), xi–xii.

29 "When Paul's Said and Done: Paul Ehrlich, famed ecologist, answers questions," *Grist*, August 13, 2004.

30 Bill McKibben, *Maybe One: A Personal and Environmental Argument for Single Child Families* (New York: Simon & Schuster, 1998).

31 James Surowiecki, "Leave No Parent Behind," *The New Yorker*, August 18, 2003.

32 Ranae J. Evenson and Robin W. Simon, "Clarifying the Relationship Between Parenthood and Depression," *Journal of Health and Social Behavior* 46, no. 4 (December 2005), 341–358.

33 Gabriel Marcel, "The Mystery of the Family," reprinted in *Homo Viator: Introduction to a Metaphysic of Hope*, trans. Emma Craufurd (London: Gollancz, 1951), 70–1.

34 Frank Newport, "Desire to Have Children Alive and Well in America," Gallup Poll Tuesday Briefing, August 19, 2003. Cited in Longman, *Empty Cradle*, 83.

CHAPTER NINE · IN WHOSE IMAGE SHALL WE DIE?

1 The previous two paragraphs draw heavily upon a working paper I co-drafted with Alan Rubenstein and Erica Jackson for the President's Council on Bioethics in my role as a senior consultant. Alan Rubenstein, Eric Cohen, and Erica Jackson, "The Definition of Death and the Ethics of Organ Procurement from the Deceased," staff discussion paper prepared for the President's Council on Bioethics, September 7, 2006 (http://www.bioethics.gov/background/rubenstein.html).

2 Phillip Longman, *The Empty Cradle: How Falling Birthrates Threaten World Prosperity and What to Do About It* (New York: Basic Books, 2004), 8.

3 See, for example, Richard Epstein, "Kidney Beancounters," *Wall Street Journal*, May 15, 2005. Sally Satel, "Death's Waiting List," *New York Times*, May 15, 2006.

4 See Genesis 49:1–33.

5 The account that follows relies heavily on the work of my friend and teacher Leon Kass.

6 Genesis 48:2 (JPS).

7 Genesis 48:3–4 (JPS).

8 Jacob's recognition of his own limits in fulfilling God's covenant is notably revealed in his near-final encounter with Joseph. See Leon R. Kass, *The Beginning of Wisdom: Reading Genesis* (Chicago: University of Chicago Press, 2006), 638, 644.

9 Genesis 48:16 (JPS).

10 *The Beginning of Wisdom* 636–8.

11 See Genesis 18:9–15, 21:1–8, 30:1–6, 30:22–24; I Samuel 1:1–2:11.

12 See Genesis 9:1–7.

13 For this interpretation of the relationship between Socrates and the oracle, I am indebted to Leo Strauss, "Jerusalem and Athens: Some Preliminary Reflections," *Leo Strauss: Studies in Platonic Political Philosophy* (Chicago: University of Chicago Press, 1983) 147, 171.

14 Xenophon, *The Apology*, trans. H. G. Dakyns, 1998 (http://www.gutenberg.org/etext/1171).

15 *Ibid.*

16 President's Council on Bioethics, *Taking Care: Ethical Caregiving in Our Aging Society* (Washington, DC: President's Council on Bioethics, 2005), 113. Available online at http://www.bioethics.gov.

17 Xenophon, *The Apology.*

18 *Ibid.*

19 Caitrin Nicol, in a personal communication to the author, October 5, 2006.

20 See Matthew 26–27 and Luke 22:53.

21 II Chronicles 16:12 (KJV).

22 Benjamin Franklin, letter to Rev. John Lathrop, 31 May 1788, in *Franklin: Writings*, ed. J. A. Leo Lemay (New York: Library of America, 1987), 1166–7.

23 Jean-Jacques Rousseau, "Discourse on the Origin and Foundations of Inequality," *The First and Second Discourses*, trans. Roger D. and Judith R. Masters (New York: St. Martin's Press, 1964), 147.

24 *Ibid.*.

25 See Albert Camus, *The Plague*, trans. Stuart Gilbert (New York: Vintage, 1948).

26 Albert Camus, "The Myth of Sisyphus," *The Myth of Sisyphus and Other Essays*, trans. Justin O'Brien (New York: Vintage, 1955), 88–89.

27 *Ibid.*, 90.

28 Yuval Levin, "The Crisis of Everyday Life," *The New Atlantis*, no. 7 (Fall 2004/Winter 2005), 120.

29 See, for example, Paul Ramsey, *The Patient as Person: Explorations in Medical Ethics* (New Haven: Yale University Press, 2002), 59–144, 239–77.

30 *Ibid.*, 66–112.

31 Paul Ramsey, *Ethics at the Edge of Life: Medical and Legal Intersections* (New Haven: Yale University Press, 1980), 145–88.

32 See Gilbert Meilaender, "'Love's Casuistry': Paul Ramsey on Caring for the Terminally Ill," *Journal of Religious Ethics* (Fall 1991), 133.

33 Yuval Levin, "Reforming the National Institutes of Health," *The New Atlantis*, no. 16 (Spring 2007), 107–111.

INDEX

Committee on Jewish Law and Standards, 115

Condorcet, Marie Jean Antoine Nicolas de Caritat, marquis de, 17–18, 24, 26

Congress, U.S., 80

"Conquest of Space and the Stature of Man, The" (Arendt), 39

contraception. *See* birth control

cosmetic surgery, 95, 96, 106, 109, 113; cosmetic society, 51

Crick, Francis, 169 *n*

Cube and the Cathedral, The (Weigel), 127

Darwin, Charles, 16–17, 20, 30, 31, 35, 37; Darwinism, 13, 16, 25, 30–34, 40

David (biblical figure), 3, 90

Dawkins, Richard, 14, 21, 32

Declaration of Independence, 55

Dembski, William, 33

dementia, 144, 145, 147, 149, 159

Dennett, Daniel, 32

Descartes, René, 7, 15, 30, 32, 84

Ditka, Mike (Coach), 108

Dolly the sheep, 81

Dorff, Rabbi Elliott, 115–117

Down syndrome, 92, 93

Ecclesiastes, 118

Edwards, Robert, 60–61

egg donation, 107

Ehrlich, Paul, 135

Elkanah (biblical figure), 130

embryo, 60–78, 114–122; cloned, 81; destruction of, 2, 4, 23, 54, 95, 114, 143; research, 4, 12, 13, 23, 48, 50, 55, 112, 143, 147, 156; moral status of, 12, 13, 22, 93, 111–112; debate, 12, 46, 156–157; man-animal hybrid, 28–29, 51, 53, 86; three-parent, 51; androgynous, 51, 86; screening of,

91, 92–93, 106, 143; as commodity, 95, 111–112; *See also* stem cells

Empty Cradle, The (Longman), 125

enhancement, 54, 67, 76–77; genetic enhancement, 51, 54, 76–77, 92, 96; pharmacological enhancement, 51, 96; surgical enhancement, 51. *See also* anti-depressants, anti-impotence drugs, *Beyond Therapy*, cosmetic surgery, genetic engineering, memory-altering drugs

Enlightenment, 18, 127

environment, 19; environmentalism, 125, 135

equality, human, 6, 13, 54–55, 60, 64–68, 75–76, 92–93, 118, 157

eschatological realism, 103, 170 *n*

eugenics; eugenic society, 51; Chinese, 52; liberal eugenics, 55; new eugenics, 77, 92, 143

euthanasia, 127, 143, 148, 157, 159

Exodus, 117

feminism, 133–134

fetus; fetal experimentation, 73–74, 78, 86; harvesting for parts, 50, 70, 73, 77–78

fire, 38–39, 40, 84

Fewer (Wattenberg), 125

Food and Drug Administration (FDA), 47, 77

Frankenstein, 86

Franklin, Benjamin, 144, 153–154, 155, 156, 157, 158

gene therapy, 85

Genesis, 13, 40, 108, 130, 151–152

genetic engineering, 52, 80, 81, 85–88, 89, 121

Gethsemane, 152

Gilbert, Neil, 133

MTV, 95
Muller, Jerry, 100

NASCAR, 95
natality, 74, 130, 143, 144
National Academy of Sciences (NAS), 29, 32
National Institutes of Health (NIH), 158
natural disasters, 12, 27, 89
Nazi experimentation, 119
New Yorker, The, 105, 136
Nicol, Caitrin, 151
Noah (biblical figure), 108, 130, 147

organ transplantation, 62, 143, 144, 147, 155; organ harvesting, 52; organ market, 96, 106, 144

Parkinson's disease, 29, 93
Paxil, *See* anti-depressants
Peck, Ellen, 134
"Pill, the," 6, 38
"Pity the Childless Couple" (Smith), 133–134
Plutarch, 131
Polybius, 128
Population Bomb, The (Ehrlich), 135
population control, 135
population explosion, 125
preimplantation genetic diagnosis (PGD), 50, 77, 92, 93
President's Council on Bioethics (PCBE), 51–52, 114, 118, 149, 169n, 171n, 174n
Prometheus, 39
Protestant Ethic and the Spirit of Capitalism, The (Weber), 97–99, 101
Protestantism, 96–103
Psalms, 6, 22, 33, 40, 90
Putin, Vladimir, 125

Queer Eye for the Straight Guy (television show), 95

Rachel (biblical figure), 147
Ramsey, Paul, 157–158
RAND Corporation, 126
Rebekah (biblical figure), 147
Rome (ancient), 129–130
Rousseau, Jean-Jacques, 154–155

Sabbath, 27, 46, 99, 115, 117
Saletan, William, 114
Sanhedrin, 117
Sarah (biblical figure), 61, 120, 122, 147
Schatten, Gerald, 81
Science magazine, 12
Seneca, 129, 173n
September 11, 2001 terrorist attacks, 46, 55
Shakespeare, William, 3, 67, 87
Simchah, R. Meir, 117
Singer, Peter, 64
Sisyphus, 18, 144, 155–157
Slate, 114
Smith, Adam, 96, 101–103, 106, 113
Social Security, 136
Socrates, 144, 148–151, 152, 153, 156, 157
Song of Songs, 38
space exploration, 6, 21, 39, 40
Sparta, 131
stem cells 3, 5, 6, 12, 20–21, 22, 28–29, 32, 51, 55, 61, 66, 73, 96, 111, 115–116, 118
sterility, self-chosen, 37, 133, 147
Steyn, Mark, 127
Strauss, Leo, 148
Swan, The (television show), 109

taboos, 15, 23, 62, 104–105
Talmud, 119

A NOTE ON THE TYPE

IN THE SHADOW OF PROGRESS *has been set in Minion, a type designed by Robert Slimbach in 1990. An offshoot of the designer's researches during the development of Adobe Garamond, Minion hybridized the characteristics of numerous Renaissance sources into a single calligraphic hand. Unlike many early faces developed exclusively for digital typesetting, drawings for Minion were transferred to the computer early in the design phase, preserving much of the freshness of the original concept. Conceived with an eye toward overall harmony, Minion's capitals, lowercase letters, and numerals were carefully balanced to maintain a well-groomed "family" appearance—both between roman and italic and across the full range of weights. A decidedly contemporary face, Minion makes free use of the qualities Slimbach found most appealing in the types of the fifteenth and sixteenth centuries. Crisp drawing and a narrow set width make Minion an economical and easygoing book type, and even its name evokes its adaptable, affable, and almost self-effacing nature, referring as it does to a small size of type, a faithful or favored servant, and a kind of peach.*

SERIES DESIGN BY CARL W. SCARBROUGH